# ADELPHI

## PAPER·286

## CONTENTS

W0018723

# ADELPHI PAPER 286

Paul R. S. Gebhard is currently Director for Policy Planning and Regional Strategies for Counterproliferation Policy in the Office of the US Secretary of Defense. As a civil servant he has previously worked on the NATO Strategic Concept and was part of the US negotiating team for the CFE Treaty. Gebhard holds a Masters in International Affairs from Columbia University, New York, and a BA from Brown University, RI.

The author would like to thank the US Department of Defense for giving him the opportunity and support to spend a year at the IISS, and the Volkswagen Stiftung for its support to the Institute and thence to the author. Needless to say, the views expressed in this Paper are those of the author alone.

None of the debts the author incurred to friends and institutions in the writing of this piece is as deep as that he owes his wife Gwenn for her unwavering confidence in a successful outcome.

First published February 1994 by Brassey's (UK) Ltd for
The International Institute for Strategic Studies

This reprint published by Routledge
2 Park Square, Milton Park, Abingdon, Oxon OX14 4RN

For the International Institute for Strategic Studies
Arundel House, 13-15 Arundel Street, Temple Place, London WC2R 3DX
www.iiss.org

Simultaneously published in the USA and Canada by Routledge
711 Third Avenue, New York, NY 10017

*Routledge is an imprint of the Taylor & Francis Group, an informa business*

**Director:** Dr John Chipman
**Production Editor:** Rachel Neaman

ISBN 1 85753 165 5
ISSN 0567-932X

The International Institute for Strategic Studies is an independent centre for research into, and a forum for information, discussion and debate on, the problems of security, strategic change, conflict and conflict prevention, arms transfers and arms control in the modern world. The Council and Staff of the Institute are international and its membership is drawn from some 90 countries. The Institute owes no allegiance to any government, any group of governments or any political or other organisation. The IISS takes a broad approach to security, and in its research and debate endeavours to develop, test and publicise the ideas and concepts underlying security issues.

The Institute's publications are designed to meet the needs of a wider audience than its own membership and are available on subscription, by mail order and in good bookshops.

# THE UNITED STATES AND EUROPEAN SECURITY

## INTRODUCTION

The security relationship between the United States and Western Europe is currently being sustained by the considerable momentum it gathered during the Cold War. The diplomatic and military cooperation that developed and the institutional structures erected to enshrine it, such as the North Atlantic Treaty Organisation (NATO), have allowed the relationship to weather infrequent yet severe storms.

The complex link between the security of Western Europe and the United States was a direct response to the Cold War. Shared values underpin the transatlantic relationship, but this alone cannot account for the enormous political and economic investment by both sides of the Atlantic in NATO's diplomatic and military structures. The level of cooperation on security and defence issues would not have come about if the threat from the Soviet Union had not existed. Now that the Soviet Union has disappeared and the Russian Army has left Eastern Europe, what are the prospects for the transatlantic security relationship and NATO in particular?

For the relationship to remain relevant it must continue to be the best tool for defending the primary interests of its members. Europe, West and East, is no longer the United States' most threatened national interest, nor will a greater mandate within Europe bring NATO greater relevance. Developments in Russia, the Middle East/Persian Gulf and North-east Asia hold greater, more credible threats for US territorial security, its allies and its national interests. At a time when the United States is cutting its defence budget and its military structure sharply, it may decide to withdraw its forces from Europe to maximise the effectiveness of its shrinking force structure for unilateral actions in regions other than Europe. However, if the US is not to act alone in defending its interests, the partners it can work with are mostly in Western Europe and Britain, France and Germany rely for their security on US engagement in Europe.

This paper argues that to address both US vital interests and those of the Western European states, NATO should become an alliance of common interests without geographic or functional limits on its competence. For example, if the United Nations (UN) asks for support for its mandate in Korea, NATO should send its forces. While the shared

3

values of the alliance remain, its common interests may no longer be as important or widespread as they were during the Cold War.

For now the momentum remains, but it cannot drive the transatlantic security relationship forever. In Europe, the importance of a US role in its security is widely recognised and accepted. The transatlantic relationship remains essential and relevant and needs no further justification.

In the US, its role in European security is subject to periodic review by the Executive, the Congress and academics. Americans who argue in favour of continued US engagement in Europe must identify the benefits for the nation's security. In 1992, Congress set the level for US troops in Europe at 100,000, 33% below the President's request. In 1993, Congress began considering legislation for the complete withdrawal of US forces unless Europe paid a greater portion of the cost of maintaining them. During the Cold War, there was domestic support in the US for the transatlantic relationship because it addressed a fundamental US national interest. Recent Congressional actions, however, demonstrate that this link is no longer obvious. If the relationship is to remain relevant to US foreign and security policy, it must address those security interests which are currently under threat.

Whether the Cold War rationale for US engagement in European security is still sufficient is discussed in Chapter I. Without the overwhelming and immediate threat posed by the Soviet Union, should the US continue to devote a huge proportion of its diplomatic and military attention to Western European security? The challenges facing the US outside Europe may best be addressed in partnership with those Western European states who are no longer at great military risk. Thus any new rationale for US engagement in Europe may have to demonstrate how the relationship directly helps the United States address its security interests in Russia, the Middle East/Persian Gulf and Korea. To what degree is a US commitment to multilateralism necessary for Western European participation in actions outside Western Europe? This emphasis may conflict with the preferred policy in the US of working with its Western European allies when possible, but addressing its interests alone when necessary.

Chapter II analyses the contribution key Western European states – Britain, France and Germany – seek from the US for European security. Are they asking the US to play a greater role as they begin to view their security more broadly to include economic, political and social issues as well as military questions? How are the Europeans offering to support US security interests outside their region, thereby encouraging US acceptance of a security role within it? European states are also currently cutting their defence budgets which will dramatically decrease their ability to contribute to their own security at home, much less to security operations beyond it.

Chapter III analyses the choices US military planners are making to increase the effectiveness of a smaller force structure. These affect whether the US relies on unilateral options or multilateral coalitions to defend its vital national interests. US military planning has paralleled the shift in political and diplomatic attention to national interests at risk beyond Western Europe, including Russia, the Middle East/Persian Gulf and Korea. Will a broader set of missions and a smaller force structure push US policy towards greater reliance on unilateral initiatives, or towards greater emphasis on ensuring European support for US-led multilateral actions?

Chapter IV analyses the effect of recent cuts in the US defence budget on whether the US will base forces in Europe or at home. As a general rule, a military base brings federal money to the region. A smaller budget and force structure means the closure of bases and the loss of federal funds to the local economy. The advantages of a US military presence overseas, such as deterrence of conflicts and influence over European affairs, must be weighed against the added cost of overseas basing and the perceived trade-off of dollars and jobs in the United States or in Europe.

Chapter V looks at options for rebuilding the transatlantic relationship and their institutional implications that must be considered even after the achievements of the 1994 NATO summit. What are the options for remaking or reinvigorating NATO? In which state is its relevance most in doubt and how can it be made relevant again? Is NATO still the best way to continue the US role in European security and for the US to earn European support for its own security concerns? NATO's January 1994 summit endorsed US proposals for a Partnership for Peace with non-NATO European states, for Combined Joint Task Forces to be used in non-Article V operations, and for greater allied efforts against the proliferation of weapons of mass destruction. These US initiatives share the basic characteristic of expanding the Alliance's area of concern for security matters without addressing the basic issue: rather than being bounded by the geography of Europe should NATO be bounded only by the common interests of its members? A transatlantic security relationship that addresses common interests represented by NATO would be quite different, but would remain vital and serve the security interests of Europe as well.

The US and its Western European allies may be unable to agree on a set of common interests. If so, the transatlantic security relationship of on-going consultations and cooperation embodied in NATO will be at an end. NATO will not be disbanded; it will just become moribund. Even if they do agree, the analyses and policy prescriptions may increasingly diverge. Until such a decision is made, however, it is illogical to assume that the tools developed during the Cold War by the US and Western Europe to protect their security interests are by definition useless now that it has ended. The United States is becoming more integrated into the economic and political affairs of the world, irrespective of the end of the Cold War. A natural complement to such a trend would be the strengthening of its strategic ties with the other major world powers in Europe and Japan. Before habits of cooperation are forgotten and institutions stagnate, the usefulness of these Cold War tools for the post-Cold War US–European security agenda should be evaluated. Such an evaluation must be preceded by a thorough analysis of the current security interests, goals and requirements of both the US and Western Europe. The marginal cost of maintenance will invariably be less than attempting to rebuild alliances in times of crisis. For it to be maintained, each member of the transatlantic alliance can and must contribute to the security concerns of all. At the centre of this sharing of resources lies the US role in European security.

# I. THE STRATEGIC RATIONALE FOR US ENGAGEMENT IN EUROPEAN SECURITY

As the global strategic environment has changed with the end of the Cold War, so has the rationale for US policy in Europe. If Western Europe is no longer the US interest under the greatest threat, will the United States in future defend its interests around the world with multilateral initiatives, or with unilateral responses and *ad hoc* coalitions? A sustainable rationale for US engagement in Europe must be an integral part of a wider, global US strategy. This chapter seeks to determine the most sustainable rationale for continued US engagement in European security.

## The Regional Strategy: US strategic planning post-Cold War

From a European perspective, the United States appears to be retreating, as evidenced by the state of US–European relations, the decline of the US military presence in Europe and its reluctance to send troops to Bosnia-Herzegovina. In mid-1993, off-the-record remarks by a high-ranking official at the Department of State on the declining willingness of the US to lead the transatlantic alliance has strengthened this perception.[1] Equating a lack of US engagement in Europe with a US retreat from international politics, however, falsely equates Europe with the world. Far from retreating from the world, the US is shifting its diplomatic and military resources to those regions where its interests are at greater risk than in Western Europe.

The diminished economic and military potential of Russia dramatically lessened the possibility of large-scale war in Central Europe at short notice. The possibility of a single, undemocratic state controlling Europe and threatening the US also decreased. Future conflicts are likely to be smaller in absolute terms, more unpredictable and perhaps more frequent than those planned for during the Cold War. The Bush administration in 1990 codified this thinking in a new National Security Strategy, whose centrepiece was a Regional Strategy for US defence policy.

In the Regional Strategy the new strategic landscape was assessed in a way that reflected the thinking of the mainstream US foreign-policy community. Statements by the current President and his top officials indicate that the Clinton adminstration has accepted its premises and logic[2] and it is becoming part of the accepted, non-partisan framework for thinking about US security and defence policy.

The Strategy's goals are best quoted rather than summarised:

– Our most fundamental goal is to deter or defeat attack from whatever source, against the United States, its citizens and forces, and to honour our historic and treaty commitments.

– The second goal is to strengthen and extend the system of defense arrangements that binds democratic and like-minded nations together in common defense against aggression, builds habits of cooperation, avoids the renationalization of security policies, and provides security at lower costs and with lower risks for all. Our preference is for a collective response to preclude threats or, if necessary, to deal with them is a key feature of our Regional Defense Strategy.

– The third goal is to preclude any hostile power from dominating a region critical to our interests, and also thereby to strengthen the barriers against the reemergence of a global threat to the interests of the United States and our allies. These regions include Europe, East Asia, the Middle East/Persian Gulf, and Latin America. Consolidated, non-democratic control of the resources of such a critical region could generate a significant threat to our security.

– The fourth goal is to help preclude conflict by reducing sources of regional instability and to limit violence should conflict occur. Within the broader national security policy of encouraging the spread and consolidation of democratic governments and open economic systems, the Defense Department furthers these ends through efforts to counter terrorism, drug trafficking, and other threats to internal democratic order; assistance to peacekeeping efforts; the provision of humanitarian and security assistance; limits on the spread of militarily significant technology, particularly the proliferation of weapons of mass destruction along with the means to deliver them; and the use of defense-to-defense contacts to assist in strengthening civil–military institutions and encourage reductions in the economic burden of military spending.[3]

The Regional Strategy recognises that 'a collective effort will not always be timely and, in the absence of US leadership, may not gel'. 'Where the stakes so merit, [the United States] must have forces ready to protect our critical interests'.[4]

During the presidential campaign, Clinton embraced this tenet in a speech to the UN Association of the United States:

Effective multilateral approaches are in the American national interest. By sharing burdens with other countries . . . we can save both lives and money . . . When multilateral approaches are unavailable or unwise, a Clinton administration would not hesitate to take the steps necessary to protect our global interests and principles. A strong UN cannot substitute for a strong national defense and foreign policy. We will act together when we can; on our own when we must.[5]

Former Secretary of Defense Les Aspin's 'Bottom-Up Review' of defence policy for the new administration also preserved this key principle of the Strategy.[6]

### Risks and threats
Secretary Christopher in his confirmation testimony outlined the regional areas of concern to the new administration.[7]

DENUCLEARISATION AND REFORM IN THE FORMER USSR
The first priority for US policy is its most immediate and potential threat: the disposition of nuclear weapons in Russia. The Russian Army may be retreating from Eastern Europe, but Russia's intercontinental missile force remains deployed and a daily risk for the United States. Effective civilian control over these weapons is essential to ensure their safety, block their potential sale and prevent their unauthorised transfer. If the other nuclear republics of the former USSR – Ukraine, Kazakhstan and Belarus – are able to target and launch the systems, this will increase the risk to the security of the United States. Its main risk, however, stems from the sale or transfer of these weapons rather than from their immediate use.

Reform in Russia is essential for US military and economic security. As Secretary Aspin once warned: 'If, for example, reform were to fail in Russia in favour of dictatorship or anarchy, the risks and costs to the West and to Russia's Eastern European neighbours become very, very grave indeed. It would shatter the prospect for partnership, place the evolving European security institutions under stress, dramatically increase the risks of conflict in the East, and impose new demands on our own – meaning the United States – defense spending'.[8]

EUROPE
One US goal is to prevent long-term threats from developing in Europe. If the immense human and economic resources in Eurasia were

controlled by a power hostile to the United States, it would be at great risk. The major strategic change caused by the end of the Cold War is the complete lack of such a threat in the foreseeable future. Neither Germany nor Russia is likely to pose a challenge, nor are there any new aspirants.

## ASIA

The demise of the Soviet Union is not a sufficient reason for the US to give up its long-standing security alliances with Japan, South Korea, Australia, Thailand and the Philippines. South Korea and Japan are threatened by North Korean conventional forces and nuclear developments, and are therefore in need of US support. The United States is also concerned about China's export of sensitive military technology that could be used for producing weapons of mass destruction or their delivery. Thus it sees both its commitment to the security of its allies in the region and the continued presence of its military forces as essential for regional stability. More fundamentally, these alliances, like those with Europe, are the strategic links the US uses and will continue to use to further its national objectives by means of its foreign policy.

## MIDDLE EAST/PERSIAN GULF

The two primary US interests in the Middle East and Persian Gulf are support for Israel and access to oil. Israel, like South Korea, is a US ally under direct threat from external military forces. The United States is also developing strong relationships with other states in the region, notably Egypt and the Gulf States. Both Iran and Iraq have threatened US security interests there over the past 15 years. Historical sensitivities among the states in the region to a Western military presence will make it difficult for the US to base more than a small force in the region. Instead, its security commitments there will have to be supported by projecting forces either from Europe or from home.

## LATIN AMERICA

The United States has important national interests in Latin America but, like its interests in Europe, they are not at great risk. Financial and military support from the Soviet Union for regimes and guerrilla groups opposed to the US is gone, although it continues to exert political and economic pressure to isolate the Castro regime in Cuba.

## PROLIFERATION OF WEAPONS OF MASS DESTRUCTION

With the exception of the nuclear weapons in the former Soviet Union, preventing the proliferation of weapons of mass destruction is the

10

United States' most important security concern. Only such weapons can seriously threaten its territorial security. Their proliferation also raises the stakes of regional conflicts. Deterring regional aggressors who possess these weapons from threatening US interests becomes more dangerous and unpredictable; and defending US interests in the region becomes a more difficult political decision. Therefore, the viability and credibility of the Regional Strategy depend to a great extent on the United States continuing to prevent proliferation and enhancing its ability to protect its forces where proliferation occurs.

Proliferation is an issue that cannot be addressed unilaterally. The technology for producing nuclear, chemical and biological weapons is widely available among US friends, allies and former adversaries and hostile states alike who also share the security risk from proliferation. Secretary Christopher was blunt in his meetings with European Foreign Ministers in Luxembourg, June 1993: 'the most urgent arms control issue of the 1990s – proliferation . . . strong, collective action by the United States and Europe is required to deal with the proliferation of weapons of mass destruction, missiles for their delivery, and sophisticated conventional arms and dual-use technologies'.[9]

## INTERNATIONAL BEHAVIOUR AND NORMS OF CONDUCT
In the states and regions where the US has little or no strategic interest, instead of being driven by strategic necessity, US policy-makers can decide whether or not to be involved in a given issue. How the Clinton administration will use this freedom has not yet become clear, although further democratic reform and humanitarian relief are two areas mentioned by the President and top officials as requiring greater attention than they were given under the Bush administration.

Once US troops are committed to a relief effort or humanitarian intervention the flexibility of US policy decreases dramatically and the safety of the troops involved becomes a top priority. In early 1994, US troops in the Persian Gulf and Somalia were undertaking delicate missions under conditions bordering at times on open combat. As the administration's policy on Bosnia makes clear, Clinton knows how his willingness to support humanitarian goals can disappear once the issue of US troop involvement is raised.

### US security policy towards Europe
The implications of the Regional Strategy for US policy on Western European security are fundamental. The logic of the Regional Strategy is that the United States sees Western European security as a distinct regional issue where the US has interests at risk. These interests have

not changed with the end of the Cold War, but the threat to them is almost gone. Yet the threat posed to its other interests around the world has not declined to the same degree.

The relative decrease in the threat to US interests in Western Europe has prompted the United States to reconsider how to allocate its diplomatic and military resources to protect its other interests around the world. Western Europe has now become a region, like many others, in which the US must determine the rationale for its policies, and the degree and the means it will use to defend them.

At the same time, however, the Regional Strategy relies strongly on alliances to bind like-minded democratic states together against aggressors. Is Western Europe just one region among many where US interests are at risk? Or is it home to the most likely and capable allies of the US, allies it will need in addressing security outside Western Europe as well? The answer to this frames the purpose of US engagement in Western European security.

## COLD WAR RATIONALE FOR US POLICIES IN EUROPE

US policies in Western Europe are rooted both in a perception of shared interests and common values, and US national interests. Americans realise that Western Europe's human and economic resources, if controlled by an anti-democratic hegemon, could pose a direct threat to US territory and interests around the world. Except for that posed by the nuclear force of the former Soviet Union, the threat from hegemonic control of Europe is the most important security concern of the United States.

Only in Western Europe has the US participated in three major wars this century in defence of its interests. In the First World War, the Second World War and the Cold War it fought against states seeking hegemonic control of Western Europe. In the first two instances, the threat came from within the region: from an uncontrolled breakdown of international relations in the First World War; and from a planned assault on the rest of the region in the Second World War. In the third instance, during the Cold War, a state sought hegemony over Western Europe from outside the region.

Because of the global dimension of the Second World War, the universalist communist goals of the Soviet Union and the idea of shared transatlantic values, during the Cold War the US viewed its role in Western European security as directly connected to its own national interests and global security policies. This view persisted in spite of evidence – Suez, Vietnam, Israel – that European allies did not always share or wish to support the US agenda in other regions.

This formed the basis for the three-part rationale for US engagement in European security. First, Western Europe needed to be protected from Soviet aggression. Second, it needed to be protected from sources of instability and insecurity within the region. Third, US engagement in Western Europe was essential to US security worldwide and could help build transatlantic coalitions for joint diplomatic and military action outside Western Europe.

These three rationales were not of equal importance. The immediate and overwhelming threat to Western Europe from the Soviet Union gave the first rationale alone sufficient weight to justify US engagement in the region's security. Collective defence to match the numerically superior Soviet force was the by-word of US–European strategic thinking. The other rationales for US engagement – addressing sources of instability and insecurity arising from within Western Europe and addressing security concerns in the Middle East and elsewhere – came usually to be considered of lesser importance. Moreover, the secondary and tertiary arguments did not elicit the same need for collective action in the face of a communal threat that defined the primary rationale for US engagement.

The anomaly of the US commitment to the collective defence of Western Europe after 150 years of avoiding 'entangling' alliances is well known. Only under special circumstances was the US willing to break with past practice and pursue its security goals through multilateral commitments. These commitments are enshrined in Article 5 of the North Atlantic Treaty signed in Washington on 4 April 1949:

> The parties agree that an armed attack against one or more of them in Europe or North America shall be considered an attack against them all, and consequently they agree that, if such an armed attack occurs, each of them in exercise of the right of individual or collective self-defence recognized by Article 51 of the Charter of the United Nations, will assist the Party or Parties so attacked by taking forthwith, individually, and in concert with the other Parties, such action as it deems necessary, including the use of armed force, to restore and maintain the security of the North Atlantic area.[10]

In this Treaty, the US accepted a degree of diplomatic and military integration with its Western European allies that it did not accept in other post-Second World War arrangements. In the Far East, its alliances with South Korea and Japan were not multilateral to the same degree as in Europe. Because of the histories of South Korea and Japan, the US could act in a more unilateral manner in its dealings with them. In Europe, by contrast, not only did a shared threat overwhelm each ally alone, but also a shared history, shared values and a common vision. Thus, the US was confident of political and military cooperation in the common goal of protecting Western European security. There was little chance that one of the allies would opt out of the coalition at the last moment exposing the US and its troops to unexpected risks.

If the US had few fears about European steadfastness, the Western European allies continually questioned American willingness to put US territory at risk for Europe. The Western Europeans felt that they, unlike the Americans, had little choice but to go along with the US (see Chapter II).

The United States' commitment to multilateral diplomacy in Western Europe did not affect its preference for unilateral measures elsewhere. This preference was reinforced by several failures to elicit European support for US policies outside Europe. European support for *Operation Desert Storm* in 1991 contrasts with its failure to provide US transit rights to support Israel in 1973, or to allow US overflight of continental Europe from Britain to bomb Libya in 1986. In these cases the European allies could be said to have pursued their own unilateral policy of not becoming involved. These incidents and others caused some Europeans to emphasise regional security issues in discussing the US role in Europe and play down the possibilities for global coalitions. In spite of complaints about unreliable European allies, the US commitment to multilateral diplomacy and military integration in pursuit of its security interests in Western Europe remained. This comfortable accommodation between the US and Western Europe lasted until the end of the Cold War.

It is unlikely that the United States will be able credibly to maintain a multilateral commitment to Western European security and a unilateral approach elsewhere. US policy-makers in future will need to determine whether the multilateral approach of transatlantic relations can be extended and expanded.

## POST-COLD WAR US POLICIES TOWARDS EUROPE

The Western European allies are the primary potential source of multilateral diplomatic and military support for US regional security policies around the world. In future, the US role in European security will be as much defined by how the security of Western Europe (and the policies of Europe's leaders) fits into US security policies in other regions as by the contribution the United States can make to Western European security. This point is important when analysing the two remaining rationales for US engagement in European security – addressing sources of instability within Western Europe, and building a coalition with Western Europe for diplomatic and military actions beyond the region. The primary rationale of the Cold War period – to protect Western Europe from the Soviet threat – provided an umbrella for other US goals in Europe. However, neither of the two remaining rationales is a lesser case of the other.

## ADDRESSING INTERNAL SOURCES OF INSTABILITY

The US has a strong interest in addressing sources of instability in Western Europe. Its intervention in two world wars should remind Americans of the cost of uncontrolled internal instabilities in Western Europe. The diplomatic participation and military presence of the US in Western Europe has helped to reassure states about the reliability and predictability of their neighbours. For Germany in particular, the US has supported a multilateral framework, NATO, in which the Germans could reliably invest their national security. The success of this multilateral approach allowed the Germans and others to forgo a more national agenda. US engagement in European security has also provided a certain degree of stability within which the European Community (EC) could develop to the point where it could make its own successful contribution towards ending internecine conflict in Europe.

It is not in the interest of the United States to neglect the stability of Western Europe after working during the Cold War to secure it from external aggression. It is not clear, however, what risks Western Europe faces from internal instabilities, or whether US engagement would be necessary to counter such instabilities. Western European policy-makers remain vague about the possible sources of internal instabilities. Britain, France and Germany have different views about future threats, but agree on the US policies necessary to avoid instabilities spiralling out of control: it must remain engaged in European security as it did during the Cold War – continuity in its commitment to Europe is essential. US commitment to multilateral diplomacy and military integration is also essential to avoid a return to national and nationalistic approaches to security and defence in Western Europe.[11]

A transatlantic relationship based primarily on balancing internal Western European instabilities would be narrow, inward-looking and short-lived. The policy would come under intense criticism, not least from the US Congress, for merely continuing US support for Western European security with no reciprocal assistance for US security interests elsewhere. The United States is looking for partners with which, first and foremost, to address issues relating to Russian security and progress and which will be ready to engage with the US when security issues arise around the world. At a time when other US interests remain under threat and resources are severely constrained, it will want to extend the rationale for its engagement with Western Europe to defending these other interests as well.

## BUILDING COALITIONS: MULTILATERALISM REVISITED

US defence spending is decreasing and the US force structure is correspondingly becoming smaller. Instead of attempting to act alone, why should the US not build a reliable multilateral alliance as it did during the Cold War in NATO? A continuing US commitment to multilateralism will be essential if it is to form a new partnership with the allies to address important security issues outside Western Europe. The Regional Strategy emphasises the importance of alliances and the contributions the allies might make, but stops short of endorsing multilateralism as the only way to secure US interests. Whether the United States approaches the Regional Strategy unilaterally rather than multilaterally will have decisive implications for its role in European security.

The security issues outlined in the Regional Strategy may not provide a sufficient rationale for the US to continue its multilateral commitment to Western European security. None of the security issues in the Strategy – Russian nuclear weapons, instability in Eastern Europe, threats to Israel and South Korea, or trouble in the Persian Gulf – require the United States and Western Europe to plan for an immediate collective defence of their national territories. Nevertheless, addressing these security threats is not optional for either the US or its allies. There will be a certain luxury of choice for political leaders over why and how policies should be developed, and security policies will be subject to debate, as happened before the Gulf War.

The Soviet threat to Western Europe virtually guaranteed that the allies would meet their multilateral commitments. As the threats identified in the Regional Strategy do not have the same consequences for Western European security as did the threat from the USSR, the new multilateral commitments of the Western Europeans are not as certain or predictable as the old ones. Because Western European leaders will have greater freedom of choice, they may sometimes choose to block or opt out of a multilateral initiative.

The US remains wary of multilateral commitments because of this unevenness and unpredictability. Without the virtual certainty of European cooperation on vital issues, the US will be reluctant to commit itself to multilateral structures where initiatives might be blocked. Maintaining a high degree of freedom for unilateral action may prevail over efforts to pursue a multilateral approach with Western Europe on global issues.

Congress has used the deadlock of US–European policy in Bosnia to cast doubt on US commitments to multilateral diplomacy and military action. In the view of some members of Congress, it was the US commitment to multilateralism that allowed the allies to block useful action in Bosnia. Senator Lugar has said: 'The "new" policy [adopted by the United States and Europe for Bosnia] reflected the abandonment of United States leadership and decisiveness in favour of "multilateralism" and the desire to pursue consensus'.[12] Other commentators have been sharper. 'Whatever its merits, Washington's Bosnia policy adds to the spreading impression that multilateralism could serve as a cloak, or an excuse, for an American retreat from the expenses and headaches of exercising world leadership'.[13]

This critique of multilateralism based on US policy in Bosnia is, however, ill-founded. It indicates a failure to understand the link between the willingness of a state to assume risks and the ability of that state to lead others. In the Persian Gulf, the US led the multilateral coalition because it put the greatest number of resources and military capabilities at risk. The same was also true of Somalia. In both cases, leadership of a multilateral coalition came from willingness to take the greatest risk.

In Bosnia, the US was not willing to assume the greatest degree of risk or even to share risk equally with its allies. Without an agreed peace plan, it was unwilling to commit any ground forces, much less to have the largest contingent. This policy protected US troops, but did not provide the US with any grounds for leadership of the coalition. When it did try to exercise leadership – to lift the arms embargo or make air strikes – it had no leverage with the allies. Had US initiatives been accepted, they would have created greater risks for the large contingent of allied troops on the ground than for the few US forces involved. Thus following the lead of other states in a multilateral alliance is the only course available for those unable or unwilling to assume the largest portion of the burden.

### Implications
The Soviet threat helped to create the diplomatic and military consensus for strong multilateral commitments to security. If the end of the Soviet Union has produced a less dangerous world in general, and in

Europe in particular, there is less need to work with other states, not more, and arguments that the US must work with its allies even more closely than before will be met with scepticism. The burden of proof will thus be on those arguing for greater cooperation and a continuing multilateral approach. The end of the Cold War has therefore not lessened, and may have reinforced, the American preference to act with its allies when possible, but to defend its interests alone if necessary.

The primary reason for the US to integrate its diplomatic and military policies towards Western European security into a multilateral institution no longer applies. If the US is to remain engaged in European security, the rationale will have to include the two remaining strategic objectives of the US in Europe: protection of Western Europe from internal threats; and US–European cooperation for addressing non-European security issues. America's leaders now need to determine whether these ends and the means necessary to achieve them are supported both at home and in Europe.

Previously, the US was content to use the transatlantic relationship to protect its interests only within the limits of Western Europe. In part, this policy was designed to protect US interests worldwide from global conflict that might begin in Europe. Without the Soviet Union, however, this link between Western European security and global security is much weaker.

Maintaining geographic limits on the transatlantic relationship would be the policy of least resistance. The allies seek continuity in the US role and do not want to be asked to participate in policies and operations outside the European continent for which they have neither significant capability nor the money. This is the choice of those not interested in giving the sustained attention to the transatlantic security relationship necessary to gain Western European agreement on global multilateralism.

Yet maintaining geographic limits on the transatlantic relationship risks diminishing the relevance of the relationship further. NATO's new Partnership for Peace (see Chapter V) modifies NATO geography but remains limited to Europe in a way that the second new initiative, Combined Joint Task Forces, does not. A complete breakdown of the relationship would neither serve the US security objectives in Europe nor support the US Regional Strategy. Nor would this benefit Europe. Nevertheless, it is not impossible for transatlantic relations to break down over trade or burden-sharing. The unintended outcome of such a break might be a statement of ambivalence by European leaders about a US presence and US withdrawal as a result.

To avoid this the transatlantic relationship must be recast to serve a widened set of shared European and American interests. These interests include, but are not limited to, Western Europe. Success would depend on two factors. First, a European realisation that US interest in its security cannot be assumed, but can be influenced by European interest in US security concerns. Second, a realisation in the United States that a more consistent, multilateral approach to security policy will help it make the best use of European resources for US security. Supplementing unilateral policies with *ad hoc* coalitions is not an effective option. Where *ad hoc* coalitions have been most effective, they have been grounded on a long-standing multilateral commitment – NATO.

The elimination of the Soviet threat to Western Europe is thus testing the breadth and depth of the common values and shared interests of the US and Western Europe. Whether those values and interests apply only to the defence of Western Europe or hold more generally for security issues around the world will greatly influence the role the US will play in European security in the future.

## II. THE EFFECT OF WESTERN EUROPEAN SECURITY POLICIES

British, French and German security and defence policies will strongly influence US willingness to contribute to European security, and they now need to decide what role they wish the US to play in their security. Should this role expand geographically across the continent? As economic, social and political issues become more important for Europe's security, should the US participate in fora governing European policy on these issues as it has in fora governing defence questions? More specifically, Europe needs to determine the extent of its ambitions to set and implement security policies for the transatlantic relationship. Should NATO be limited to European security issues, or should it address issues of common concern wherever they occur? Once policy is set, will Europe develop the military capabilities necessary to support its policies? Western Europe must now choose security policies and develop military capabilities that either allow it to become independent of the US, or provide incentives for the US to participate in European security. The danger is that Europe will select policies and develop capabilities that provide neither independence from, nor incentives for, US participation.

### European views on the US role in European security

As the United States played a major role in European security during the Cold War, it follows that changes in the threat to European security should lead to changes in this role. Surprisingly, this is not the case. There is a strong consensus among the primary Western European states that the US should continue to play a major role in their security.

There is no consensus among key states, however, on the threats to European security. Different states want the US to continue its role for their own reasons, according to their views of the risks to European security. The apparent illogic of the strong, powerfully voiced consensus on the need for the US role and the absence of a consensus on why such a role is needed is not new to Europe. This chapter thus begins where European thinking does, with the need for a US role, and not with perceptions of the threat.

### THE UK VIEW

Most British leaders take for granted that any diminution of the US role in European security is a loss for European security as a whole. Foreign Secretary Douglas Hurd argues that Europe will always be more secure when the US plays a central role.[1] This argument appears to hold whatever the threat to European security and whatever the condition of European-based defence capabilities. In practical terms, the US provides essential military capabilities that Europe would find too expensive to field on its own. As Hurd has said, 'No European democratic politician could hope to generate the support needed to replace the resources which the United States brings to our collective security, and the European Union without the underpinning of the Alliance would be a much less stable place'.[2] A US military presence in Europe, and reinforcement if necessary, is thus strongly supported by the British.

In addition, Britain is loathe to give up its 'special relationship' with the United States because if the US plays a strong role in European security the special relationship will enhance the British position in Europe. The UK can also avoid putting all its diplomatic energy into the EC as long as the US plays a strong role in European security. The British position on the US role in Europe thus owes much to its view of its own place within Europe.

THE FRENCH VIEW

The French view of the US role in European security changed sharply during spring 1993. Immediately after the Cold War, French leaders had strongly asserted that European institutions, the EC and the Western European Union (WEU), could set European security policy and that Europe should no longer accept US leadership in doing so.

According to this line of thinking, America's strong role in European security was only necessary to cope with the overwhelming military threat posed by Soviet conventional and nuclear forces. According to Prime Minister Pierre Bérégovoy in autumn 1992, 'From now on, with Germany's reunification, the disappearance of the Soviets, a Europe which has wiped out its divisions is seeking the means to organise itself to maintain peace within its borders'.[3] His Foreign Minister, M. Dumas, was even more specific. In future, the EC should set the common defence policy for Europe, and the 'WEU must be the organ implementing that common defence'.[4] The US role in European security was still essential, but only in the way that reserve forces are seen as essential to the defence of a nation. The US, like the reserves, would only be called upon when necessary. European security on a day-to-day basis would be handled by Europe.

French arguments for European control of its security resonated well during the years between the fall of the Berlin Wall and the Maastricht Treaty. The political momentum seemed to favour Europe and European-based institutions as the new centres of economic, military and political power. Although never stated explicitly by French leaders, the shift in decision-making on security policy from the US towards Europe was not seen as a single event. Starting from the end of the Cold War, Europe would progressively wrest greater and greater control from the United States. In this atmosphere, a French policy to diminish US influence in Europe became an issue of European independence from the United States. France's ability to define issues of French national policy as litmus tests of support for European Union made it difficult for the British and Germans to object as neither had the credibility on European issues to withstand repeated assaults on their support for 'Europe'.

Because France was able to form the policy of the European Community in its own image, the change in French policy in 1993 had broad ramifications. The French leadership would not give up Europe's prerogative to set European security policy. Neither, however, were French leaders still talking about marginalising US influence.[5] French Foreign Minister Alain Juppé said in an interview with *Le Figaro*: 'The Alliance [with the United States] must evolve. But there must still be an alliance: the threat is no longer the same, but dangers persist'.[6] In the same vein, French Defence Minister François Leotard said: 'France considers that today's threats are different, and that NATO is adapting to these new threats. We should pursue this evolution of NATO and of France, pragmatically and without dogmatism, and France's attitude is open [on participating in NATO's Military Committee and in planning for humanitarian actions]'.[7] French officials acknowledged that the Community and its affiliated bodies, such as the WEU, would never displace NATO and the US contribution to European security. Juppé was disarmingly clear on the limits of the WEU to his European colleagues at a WEU ministerial session. 'Realism, at the same time, because the WEU is clearly not cut out to do everything by itself in the security field. The Yugoslav crisis shows that we can have recourse to the capabilities of the WEU or NATO, in a complementary way, depending on the envisaged operation, and quite obviously depending on the origin of the forces engaged'.[8]

The Balladur initiative at the EC's Copenhagen Summit in June 1993, now adopted as the European Union's initiative, reflects the new compromise in French thinking. Balladur's plan modified an earlier idea President Mitterrand had for addressing European security. Mitterrand had called for a European confederation that was pan-European, but not transatlantic. Balladur's plan also focused on bringing stability to Eastern and Central Europe, but highlighted the participation of the United States and Canada. By including the US in his plan, Balladur acknowledged the inability of the Community credibly to address European security issues alone.[9]

The change of government in Paris and the situation in Bosnia had great influence on French policy. The victory in spring 1993 of the conservative right resulted in a shift of influence away from the Quai D'Orsay. The constitution of the Fifth Republic does not grant responsibility for foreign and security policy fully to the President or Prime Minister,[10] thus as long as one political party controls both positions, there is no political tension at the top. Without political tension, the bureaucratic institutions of government retain their day-to-day control over policy. The Quai has traditionally exercised control over French policy towards NATO because of its role as the protector of Gaullist particularism and separation from NATO. When conservative Prime Minister Balladur was forced to share power with socialist President Mitterrand after the elections in spring 1993, however, the practice of making decisions in government changed. Because of divisions between the two leaders, there is a tendency to move delicate foreign-policy decisions to the top political level, away from the Quai. This shift in the locus of decision-making would benefit the conservatives, who have supported a greater degree of French cooperation with the alliance.

The poor performance of the EC in Bosnia also affected French policy on the US role in Europe. French leaders were frustrated by the EC's inability to do more than provide humanitarian relief. Foreign Minister Juppé admitted in *Le Figaro* that Bosnia could be seen as a sign of Europe's impotence and as a failure of the Community's common defence and foreign policy.[11] The French are unlikely to give up immediately their long-held fears of American hegemony in Europe and support a stronger US role. There is evidence, however, of a realisation in France that the immediate European problem is keeping the United States interested in its security.

THE GERMAN VIEW
German leaders see a strong US role as an essential part of their security policies for both Western and Eastern Europe. In the West, German leaders are committed to maintaining the multinational framework developed during the Cold War.[12] In Germany 'nationalism' is often paired with 'destructive'. Leaders avoid articulating explicitly national views of German national interests, or they deride such views as dangerous. Chancellor Kohl has been blunt: 'The spectre of nationalism is not at home only in the Balkans. Neither is Western Europe

23

forever free from the evil forces of the past, from nationalist thinking, from a fall into intolerance and chauvinism'.[13] German leaders want to continue to articulate their national interests but through the adoption of common positions in multilateral fora such as NATO.

To the East, Germany's policies are most influenced by Russia. Germany has attempted both to reassure Russia of its intentions and prepare its defences against Russia. The Germans see the US role in Europe as essential to the success of both policies. The crucial event in Germany's plan to reassure Russia of its intentions and succeed with unification occurred at the July 1990 Shelesnowodsk meeting when Kohl won Gorbachev's support for unification in exchange for a number of financial arrangements and limits on German military manpower. Russia also agreed to continued German membership in NATO.[14] Germany unofficially interprets this as the Soviet Union wanting a guarantee that Germany will remain embedded in European institutions and not develop an independent defence capability.

Germany interprets the views of the Visigrad 4 similarly. Poland, the Czech Republic, Slovakia and Hungary are seen as much more comfortable neighbours and willing partners because Germany is integrated into a strong Western multilateral framework. The US political and military presence in Germany is thus necessary to maintain NATO's credibility and to reassure Eastern states of Germany's continued commitment to peace.

Moscow may be something of a partner, but it is also a security risk for Germany presenting its only potential military threat of any size. This threat has declined since the end of the Cold War, but has also become more difficult to predict as Russia's politics and economics have grown more unstable. German leaders are reticent about commenting in public on possible political reversals in Moscow in case such speculation gives indirect support to hardliners in Moscow who believe that the West remains antagonistic to Russia. General Klaus Naumann, head of the German armed forces, has often been more direct. Noting that Russia will continue to maintain the largest land army in Europe, the largest air force and a huge inventory of nuclear weapons and that, until mid-1994, Russian troops will remain in Germany, Naumann has argued that Germany cannot handle the potential threats from Russia alone or even with the EC.[15] Germany's ability to defend its five eastern *Bundeslaender* was made more difficult in the Two-Plus-Four agreement. It agreed not to allow the stationing of non-German NATO troops in the territory of the former German Democratic Republic,[16] therefore any initial defence of eastern Germany is likely to be undertaken by the Bundeswehr alone.

Preventing the renationalisation of defence in Western Europe and reassuring the East about German intentions both depend on the presence of US forces in Germany. To realise a policy of multinational defence, the Germans want to leaven their units with US troops. During the Cold War, a large US troop presence gave the allies added assurance of the US commitment to European defence. In the post-Cold War period, the need to add credibility to US defence commitments has probably increased, not decreased.[17]

## THE EUROPEAN CONSENSUS

In summary, the British, French and Germans are unified on three points concerning the US role in European security. First, they perceive a need for this role. Second, it should be expressed through NATO. Third, a central part of the role is the maintenance of a military presence. The leaders of the European states, however, have their own reasons for believing that a strong US role in Europe is necessary. The British do not assume that a direct threat is necessary to justify a US presence and rely on the simple axiom that more is better. The French, interested in institutions, want a US role in European security to lend quiet credibility (and capabilities) to EC and WEU initiatives. The Germans want US military forces in Germany to address its security problems in the West and East. At root, what the Western European allies share is a desire for continuity in their security affairs in the face of uncertainty. It is because the threat is unknown that a US role is deemed essential.

## LIMITS ON THE US ROLE IN EUROPE

Although the Western Europeans are clear about their desire for a US role in their security, they are equally clear about the limits of US influence. Europe cannot completely isolate its decision-making processes on economic and security issues from US influence. In security matters, the US is a full member of the primary security institution in Europe, NATO; in economic matters, however, the US is an outsider to discussions in the EC. Europe will not evict the US from NATO; equally, however, the US will never become a member of the EC. Thus the allies want US participation in European discussions to remain limited to security matters.

### European actions and their effects

The US cannot exercise influence in Europe unless there is a demand for it to participate in European security. Similarly, its influence is dependent upon its willingness to participate. This willingness, in turn, is affected by several factors, including allied security policy and policy about security issues of interest to the United States.

The Western Europeans are making diplomatic and budgetary decisions that could adversely affect the US role in their security. Diplomatically, Europe has begun to use the EC and the WEU to develop its security policies without the participation of the United States. If NATO is no longer the primary decision-making body on European security issues, the US cannot participate fully in European security policies. If Europe wants the United States to participate, it should not make it more difficult for it to do so.

The cuts being made by the British, French and Germans in their defence budgets will decrease their ability to contribute to Western European security or security in any other region of the world. Clinton administration officials have voiced their concerns about this. Principal Deputy Under-Secretary of Defense, Walt Slocombe, in testimony before the US Senate, stated: 'There is a European responsibility here, too. If NATO – and Europe – is to be taken seriously in the world, Europe will need effective military forces . . . The administration has made clear its expectation that European countries maintain adequate military forces'.[18] These diplomatic and budgetary developments alone may not lead the US to end its role in Europe. Neither, however, will they improve its willingness to participate in European affairs.

EUROPEAN ACTIONS

The French and other EC members want to set the agenda for European security but without endangering US participation in it. Balladur's proposal for a conference on security in Europe includes the United States, but is designed to be run by the EC. In the past the US has questioned European proposals for changing the way security policy for Europe is made to avoid the EC or WEU presenting it with a *fait accompli*. US proposals to keep NATO the 'essential forum for consultation among its members and the venue for agreement on policies bearing on the security and defense commitments of Allies under the Washington Treaty' were in direct response to European proposals for the development of a European Security and Defence Identity in the WEU outside NATO.[19] US criticism of European institutions, however, can only be credible if European policies are unsuccessful.

To date, the EC and the WEU have had no unequivocal successes. Handling the break-up of Yugoslavia was a difficult mission for Europe and it may be that none of the solutions proposed for Bosnia-Herzegovina were viable. Whether this was the case will never be known because the EC was never able to implement its preferred policy, the UN/EC Vance–Owen plan. It did not want to invest its own resources in it and was unable to convince the United States to put its resources behind a policy which it did not help formulate. The EC failed because it could not even attempt to implement its policy, not because it failed once implemented.

The conflict in the former Yugoslavia exposes the limits of a European security policy that lacks an effective military structure. Security policies need credible and effective implementation mechanisms. While Europe may accuse the United States of choosing not to use its available capabilities, the Community can be criticised for formulating a policy that exceeds its available abilities.

The EC claimed the lead in setting Western policy at the start of the Yugoslav crisis. The United States gratefully acquiesced and did not even have an official negotiator in the two-man team of Cyrus Vance and Lord Owen. The Europeans may have thought that Vance's participation as the UN representative was sufficient to commit the US to whatever policy developed. By having a former Secretary of State on the team, they may have expected to bring the US into the negotiations without having to work with officials in Washington. This approach reflects a desire in European capitols for 'Europe' to set the political agenda without official US participation on issues of European security.

Vance and Owen came to Washington in February 1993 to convince the Clinton administration to support their peace plan for Bosnia as US political support was needed to pressure the parties in Bosnia to agree to the arrangement. Vance and Owen argued that the deal, although not completely agreed, was the best that could be crafted (implying that US participation would not have produced a better deal for the Muslims). US military support was needed to provide the largest contingent of ground forces necessary to enforce the peace. London and Paris argued that they could not add significantly to their troop deployments to Yugoslavia. (In fact, the Europeans could have deployed more forces, as they did for the 1991 Gulf War, and deployed more troops to Yugoslavia. US troops were needed to make the peace plan work because the European allies were unwilling to do much more.) The Clinton administration, however, was extremely reluctant to offer US military support. Its officials objected to the Vance–Owen agreement because it rewarded Serbian aggression and was militarily unenforceable without complete and sincere agreement by all three Bosnian parties.

The EC was asking for US political and military support for a deal in which it had taken no official part and about which it had deep reservations. Whether US participation would have produced a better agreement is irrelevant. Without its participation, the Clinton administration was not committed politically to the plan and was under no obligation to supply the largest contingent of military forces to enforce the deal.

Because of the situation in Bosnia, the EC was unable to set the agenda for European security without the full participation of the United States. The United States would have found it difficult to reject European requests to supply forces if the Community had achieved agreement on the Vance–Owen plan and been able to muster a majority of the forces required. However, the political influence and military power of the US remain essential to security arrangements in Europe and the EC and WEU's scope for meaningful developments in European security policy will consequently remain limited. The US showed by its inaction on the Vance–Owen plan that it was extremely reluctant to place American political and military resources at the disposal of the EC and the UN for use as these organisations determine.

DEFENCE BUDGET DECISIONS
With the projected defence budget cuts of Britain, France and Germany their ability to implement a plan like Vance–Owen will remain in doubt. Nor will the European allies much improve their capabilities for more challenging missions, such as participating with the US in the wider security agenda outlined in the Regional Strategy.

GERMANY
The German leadership views military forces in the post-Cold War period as most valuable for its peacetime political purposes. Its most important national goal is the quiet integration of a reunified state into Western and Eastern Europe. How the military might best contribute to this goal is Germany's most important defence issue. Creating multinational units and reassuring allies about German intentions are two examples. Because there is no longer an obvious and immediate military threat to Germany, less political importance is given to defending it against aggressors.

In Germany as elsewhere, the mission assigned to a nation's military forces primarily governs its size and capabilities. A force able to participate in multinational units without threatening Germany's neighbours neither needs to be very large nor highly ready. A force able to deter and defend against a smaller and less likely threat from the East can also be small and less than ready. This explains in part the drastic decline in German defence spending that has already taken place and is projected to continue in the future.

German military planners are reducing manpower totals and limiting the procurement of new equipment, thereby shrinking overall force size and reducing its capabilities. Official NATO figures show a 6% drop in defence spending from 1988 (57.63 billion DM) to 1992 (54.12bn DM) in real terms.[20] All new arms procurement contracts were frozen for 1993, saving about 860 million DM. Defence spending as a percentage of gross domestic product (GDP) dropped from its historical average of over 3% to 2.2% in 1992, a substantially smaller proportion of national income spent on defence than in other key European states. These continuing reductions in defence spending overall, and in procurement in particular, will slow the modernisation of Germany's armed forces. While Britain and France already have some capability to project forces abroad, the German leadership will find it difficult to redirect resources within a shrinking defence budget to develop the completely new capabilities necessary to project its forces.

The German armed forces are limited to no more than 370,000 troops after 1994 under the Two-Plus-Four agreement and the Conventional Armed Forces in Europe (CFE) 1A Treaty. This is a substantial reduction – 26% – from the Cold War levels of about 500,000 and an even deeper cut from the unified German armed forces, East and West, of about 637,000.[21] Chancellor Kohl in February 1993 called for an immediate review to determine the most appropriate troop levels and force structure. Early press reports suggest that the final total force might be between 250,000 and 300,000.[22]

If the Bundeswehr shrinks below 370,000, the length of conscription may also decrease from 12 months to no more than ten months. With 12 months of training, conscripts receive one month of leave. If the leave period is retained, the quality of conscript training will drop even further than would otherwise be expected. Even with a full ten-month conscription period, there will be serious doubts about the ability of the German forces to maintain adequate training levels and field ready units. The cohesion and readiness of units will suffer because of high personnel turnover. Ten-month conscripts will spend only a short period in their units after training and prior to discharge. These changes may call into question the ability of the German armed forces to deploy on active-duty operations.[23]

## UNITED KINGDOM

The British continue to view military power and its application as relevant to the nation's standing and its political goals. As Foreign Secretary Hurd has said, 'the fact of the matter is that if boils keep breaking out on the face of the world, our British commitments are likely to increase provided that we want to maintain our position as a medium–large power with a highly developed sense of international responsibility'.[24] This is what Hurd means when he says Britain must continue to 'punch above her weight'. UK defence budget plans belie this commitment to the use of military force for security in Europe and British interests generally. The Ministry of Defence (MoD) in its Summer 1993 White Paper shows that spending is projected to drop substantially.[25] Defence spending has already fallen by 4.5% from £16,476bn in 1988 to £15,739bn in 1992 (fixed 1985 prices, discounting inflation). Spending as a percentage of GDP has remained constant at about 4% from 1988 levels, twice the German rate, and the highest rate by far of the largest European countries.[26] However, expenditure plans for the MoD, presented on 30 November, project the defence budget falling by 14% in real terms. By 1995–96, UK defence spending will fall to no more than 3.1% of GDP.[27]

British troop levels will drop 20%, from 300,000 (1990) to 240,000 (1995), while its Army strength will drop a huge 24% from 156,000 (1990) to 119,000 (1995). UK naval strength will decline by 15% and the Airforce by 21%. The size of these cuts is important because in 1991 34% of British forces and 55% of Army troops were deployed overseas or in Northern Ireland.[28] Traditionally, overseas deployments are more expensive than home deployments.

These manpower and budget cuts are mirrored in changes in the UK military structure. Army infantry battalions will drop from 55 (1990) to 40 (1995). Royal Navy destroyers and frigates will drop from 46 (1990) to 'about 35' (1995). The Air Force will lose 20% of its offensive strike aircraft between 1990 and 1995, leaving it with 204 aircraft of which only the 112 *Tornado* GR-1s can be said to be modern.[29]

Within these overall trends towards a smaller force, the MoD is planning to enlarge or upgrade some of the special capabilities needed to project forces. As stated in the White Paper, 'We must recognise that there will be pressure on the United Kingdom to play a part in any international response [to conflicts in and around Europe], and the Government may wish to do so where appropriate. The same is true outside Europe'.[30] In addition to a recently ordered helicopter carrier, two of the UK's amphibious assault ships will be replaced. The MoD plans to modernise or replace at least half of its fleet of 60 C-130 *Hercules* transport aircraft. It also plans to procure additional helicopter support assets,[31] but it is not clear whether these investments in force-projection capabilities are worthwhile. In addition to lacking these capabilities, the basic problem the British encountered both in the Gulf War and in Bosnia was not having enough troops available to sustain an overseas operation.

The enormous proportion of UK troops deployed away from home is beginning to strain the ability of both force structure and troops to provide rotational replacements. The planning standard of the MoD is to deploy soldiers to an 'emergency' (e.g., Northern Ireland, Belize) six months out of every 30. In 1993, the Army sent troops on emergency tours for six out of every 21 months. It is meeting its commitments by pushing its soldiers harder rather than by calling-up reserves or retaining a larger force structure. Money can be saved this way in the short term. If such a pace is maintained, however, the effectiveness of the troops will decline sharply. Poor morale will lead to fewer re-enlistments, especially troubling for an all-volunteer force.[32] The options are not easy: refuse to respond to certain emergencies; or increase the size of the force structure. The MoD White Paper does not indicate what the government will do.

FRANCE

France, like Britain, uses its armed forces in combat situations to support foreign-policy objectives. Political and military leaders learned from the Gulf War how the military can support France's international status. Conversely, there has been concern among its leadership that its status may slip if the French military lacks certain capabilities. Prime Minister Balladur has requested the French MoD to produce a White Paper by autumn 1993 (still not released in early 1994) to analyse 'the new situation in which we find ourselves, the potential threats and the means our country has of ensuring its independence and survival in all circumstances. We must not take the risk of finding ourselves in the future in situations in which our armed forces were without all the means required to carry out the tasks assigned to them by the political power wherever they were needed in the world'.[33] The French government will probably maintain defence spending and troop levels to a greater degree than any of the other principal allies and also add capabilities identified during the Gulf War.

The small reductions in French defence spending indicate how seriously its leadership takes the lessons of the Gulf War and the role of the military in French foreign policy. NATO figures for French defence spending show a drop of only 0.65% from 1988 (192,420bn francs) to 1992 (191,160bn francs). Defence spending as a percentage of GDP is down to 3.4% in 1992 from 3.8% in 1988, but is still well above comparable German figures and projected UK figures. Personnel reductions in the French armed forces will be only 16% from 420,000 in 1990 to 350,000 in 1994. The French government plans to retain conscription, but is looking at ways of allowing conscripts to volunteer for overseas duty. French army manpower will drop 22% from 288,000 in 1990 to 224,000 in 1998. Again, these figures are somewhat less than British cuts, but are significantly less than the 29% reductions in the German Army.

Information on French military procurement is difficult to collect. General statements from ministers indicate that France will postpone some major arms purchases, but will go ahead with the *Mirage* 2000 fighter/bombers and increase spending on satellite/intelligence assets. Both of these were identified as key capabilities lacking during the Gulf War.[34] With personnel cuts much deeper than overall budget cuts, French defence spending on equipment per soldier is projected to increase substantially. In spite of these efforts, French spending plans are designed to fill the gaps in the modest force it deployed to the Gulf. The government, however, has no plans to be able to project a larger force than it did during the Gulf War.

EUROPEAN DEFENCE SUMMARY
In summary, the key Western European states are not increasing their ability to employ their forces in Europe or to project forces elsewhere,[35] a trend which may undermine the willingness of the United States to deploy forces in Europe to defend Europe. The US should not, however, criticise the security policies of the Europeans merely because it is not part of the inner cadre. Instead, it should set a standard for the policies of the Community: successful policies require good ideas and both the willingness and the ability to implement them.

**Between policy and implementation**
There are several ways of bridging the gap between EC/WEU security policies and the military means to implement them. The WEU states can increase their defence budgets to procure hardware and expand their force structures. The allies can restructure the forces they have in common for greater efficiency and spend the money saved on new capabilities; or, they can borrow what they need from outside sources, notably the United States. The adoption of any of these options, how-

ever, requires the WEU states to overcome significant political obstacles.

## NEW SPENDING
The Europeans are co-developing some of the capabilities that are too expensive for any of their individual defence budgets. The WEU has a new satellite centre in Torrejón, Spain, to process images from French satellites for arms-control verification. The low cost of this venture, about $44m for the first three years, may help explain its rapid approval and development.[36]

The WEU is studying deployment of a satellite system with optical photography, infra-red photography and a synthetic aperture radar for arms-control verification and crisis management. The project would enter service in the year 2000 and feed into the WEU satellite centre. Preliminary estimates of the initial cost of the system are 6bn DM, about $3.5bn.[37]

The Technology and Aerospace Committee of the WEU Assembly held a symposium in April 1993 on anti-missile defence for Europe. It concluded that the WEU should 'launch basic studies to allow Western Europe to have, in about 15 years time, an anti-missile defence system'.[38] This new interest is a reversal of earlier scepticism about US plans for missile defence. However, the rather vague, long-term timescale of 15 years makes this recommendation of little relevance in the immediate or near future. More important, neither of these proposals address the current fundamental problem of the WEU states: a shortage of ground forces available for long-term deployment and rotation.

## RESTRUCTURING FOR GREATER EFFICIENCY
In lieu of enlarging their force structures, the WEU states are trying to do more with what they have at hand. In the Petersberg Declaration of June 1992, they agreed to increase their ability to use national forces in multinational operations. Military delegations to the WEU will be augmented and given responsibility for providing military advice to the Council. The WEU established a military planning cell on 1 October 1992 to:

- prepare contingency plans for employment of forces under WEU auspices;

- prepare recommendations for the necessary command, control and communication arrangements, including standing operating procedures for headquarters which might be selected;

–   keep an updated list of units and combinations of units which might be allocated to WEU for specific operations.[39]

Finally, its members agreed to designate military units and headquarters for use in multinational WEU missions.

At the Rome meeting of the WEU Council of Ministers on 19 May 1993, several European states pledged to provide forces for use by the WEU. Germany, France, Belgium and Spain pledged jointly their units in the Eurocorps; Belgium, Britain, the Netherlands and Germany pledged their units in the Multinational Division (Centre) normally committed to NATO; and Britain and the Netherlands pledged their joint Amphibious Force, also normally committed to NATO.[40] In each case, these units are being given a third organisational affiliation, labelled WEU, to go with their current affiliations of NATO and Home Defence. No new capabilities were created through these organisational changes.

All these developments leave the basic question unanswered: how can Europe give more capability to its forces while cutting its defence budgets? The answer may be greater efficiency among the defence forces of the WEU member-states; or, it may be for the Europeans to request assistance from the United States.

The combined defence budgets for 1992 of the ten WEU states are about $150bn,[41] 60% of US defence spending in the same year. Based on aggregate defence spending, the WEU states should be able to develop some of the special military capabilities provided by the US and increase their capability to project forces. A report by RAND concludes that long-range transport assets and satellite intelligence systems would greatly enhance Europe's ability to project forces beyond Western Europe. The report estimates that these capabilities would cost the WEU between $27bn and $95bn over 25 years, depending on the degree of autonomy sought from the United States. An average cost of $1–2bn per year seems small compared to the combined annual defence budgets of the WEU. If, however, the WEU states want to accommodate these costs within their current plans for decreased defence spending, RAND estimates that spending in other areas of defence would have to be cut by between 1.5% and 4.4%.[42] The equally difficult alternative is for the WEU states to reduce the size of their 'peace dividend' and make smaller decreases in defence spending than planned.

If these states do not want to spend any more on defence than planned, they could rationalise their combined defence spending and create an integrated military structure. The NATO states have achieved some rationalisation in their defence planning by role specialisation and by holding certain assets in common. The best example of role specialisation is among its navies. The US provides the majority of the ships needed to control the sea lines of communication; Belgium and the Netherlands provide many of the minesweeping vessels for the alliance; and the alliance holds 18 E-3A airborne warning and control system (AWACS) aircraft, nominally registered in Luxembourg. Although these measures would lower national defence budgets and free money to improve the overall force, there would be difficult hurdles for national governments to overcome.

In 1993, NATO's enforcement of the 'no-fly zone' over Bosnia was jeopardised by the German nationality of the flight crews. The German leadership dithered over its national prohibition on military personnel participating in combat operations outside NATO territory. Although the AWACS planes could have been used with other crews not explicitly trained for NATO operations, the effectiveness of the operation would have decreased. This incident exposed the weakness of all commonly held assets: by withholding its participation, one state can degrade or block unanimous agreement for military action. Under the Cold War requirement for collective defence, the problem of one country opting out was assumed away. Now that states have the freedom to choose to be involved in military actions with their allies, dissenters can immobilise the coalition.

Role specialisation or procurement of special military assets to be held in common provide a strong incentive for the creation of tight military structures, such as European Command (EUCOM) and Atlantic Command (LANTCOM), US national commands within NATO. Through an integrated military structure, complementary allied capabilities can be melded into a single force. NATO's integrated military structure is one of the most important features of the alliance, making it truly greater than the sum of its parts. Through this structure, NATO is able to make more effective use of the military contributions of smaller allies who alone would be unable to field as coherent a force. The creation of the Eurocorps may be a step in this direction by the WEU. Collective funding and ownership, such as the Torrejón satellite facility, is another example. It is, however, difficult to know if the absence of the US will be sufficient for France to overcome its traditional aversion to integrated military commands.

HELP FROM THE UNITED STATES
The WEU states may neither wish to spend more money on defence, nor to divide up the missions among their national forces within an

integrated military structure. The WEU's third option, compatible with the second, is to draw on the capabilities of the United States through NATO. This 'lending library' policy would allow NATO member-states to form coalitions (such as the WEU) and borrow additional military capabilities from NATO. NATO's new Combined Joint Task Forces (CJTF), discussed in Chapter V, may open a way for the Europeans to borrow from such a 'library' of capabilities. In practice, this would usually mean US support for WEU-led operations which might benefit the US by enhancing the credibility of the WEU (if the operation were successful) and creating greater European responsibility for its own security.

US support for a 'lending library' would depend upon the type of capabilities lent. The US has traditionally provided UN operations with non-combat support capabilities, such as airlift and logistics. In Somalia, it has demonstrated that it would place forces under the operational control of a non-US commander if US forces constituted only a minority share of the principal forces engaged. The United States may provide support to WEU-led coalitions as it has to the UN, but it is highly unlikely that it will give its European allies significant combat capabilities to bridge the gap between the EC's desire to make policy and its ability to implement it.

Britain, France and Germany agree on the need for a US role in European security expressed through NATO and substantiated by the maintenance of a military presence. Missing from allied reasoning is an analysis of the US rationale for providing the forces to which Europe will be host and partner.

To the US, its troops in Europe are an integral part of the larger US force structure; as the whole is dedicated to addressing all US security concerns, so are its parts. Therefore, the standard for US forces in Europe is their ability to help protect US security interests worldwide, not only in Western Europe. During the Cold War, the mission requirements for US forces in Europe and US forces globally were well matched. In future, however, the mission requirements for US forces in Europe may not be as rigorous as the requirements for the US force structure as a whole. Rather than maintain less capable forces in Europe, the United States will want to ensure that its forces there are at least as ready and capable of engaging in missions wherever necessary as the rest of its forces worldwide.

US operational requirements for basing its forces in Europe are discussed in greater detail in Chapter III. The United States will be reluctant to maintain a military presence in Europe if the quality and size of allied contributions to multinational units declines dramatically. Europe should avoid fuelling the perception that the US is providing better quality troops for its security than they are. Similarly, if Europe is not willing to improve its ability to make a significant contribution to security operations in other regions, US willingness to accept a role in European security will decrease.

## Implications

The key Western European states want both a US security commitment to Europe through NATO, and a US military presence in Europe. The only other mechanism for an equivalent US contribution to European security would be a position within the EC based on US economic and political weight – a highly unlikely policy for Western Europe to pursue.

In the midst of political turmoil and new, unpredictable threats, Europe wants the role of the US to remain an element of continuity. The limited requirements for this US role in European security are conservative in that Europe is seeking to conserve as much of the past as possible. What the Western European approach does not take into account is the possibility that the US may not accept its offer of a role.

US leaders have maintained that their country would remain engaged in European security only as long as it was invited to do so. Europe is willing to give strong political support to a US role in its security, but US acceptance of the European invitation is not automatic. The US wants more from Western Europe than to be allowed to contribute to its security by basing forces there. Increasingly, it will seek a commitment from Western Europe to engage with the US in defence of common security goals around the world. Unfortunately, the EC's unwillingness to commit sufficient forces to support its policies, and the projected cuts in European forces will not make this commitment easy to make.

Europe is seeking greater control of its security by limiting the transatlantic relationship to a military role within strict geographic bounds. This limits American influence in Europe amidst an otherwise broadening set of political and economic concerns. By contrast, the US wants to extend the transatlantic relationship beyond Western European borders to defend common interests wherever they are threatened. Because the focus of such a transatlantic relationship would not necessarily be Western Europe, Europe would achieve its goal of greater control of its own security. If the Western Europeans are not interested in security beyond their continent, US interest in working with Western Europe will wane and the US role in its security will decline, leaving Europe to address alone whatever security problems it faces at home.

## III. US MILITARY DEPLOYMENTS IN WESTERN EUROPE

This chapter looks at how the operational requirements for US forces to support its foreign-policy goals have changed with the end of the Cold War and the decline in US defence spending. A smaller US force structure may no longer be able to reconcile the requirement to defend its interests unilaterally with a desire to support multilateral diplomacy with Europe. If US leaders continue to require the capability to act alone, the military's operational arguments against deploying US forces in Europe will be difficult to refute.

### US troops in Europe

US military planning during the Cold War gave the greatest weight to the operational requirements for deterring and defending aggression against Western Europe. This influenced the size, readiness and structure of US military deployments in Europe.

US foreign policy also benefited from stationing its forces in Europe. The presence of these forces served to promote stability in the region, as discussed in Chapter II, and US troops were also closer to hot-spots in the Middle East and to its allies in that area. These added benefits were a result not only of the presence of US forces, but of their number and capabilities relative to the allied forces, especially those of Germany. Because of the overwhelming threat to Western Europe, these added benefits did not have to be invoked as the primary justification for US military deployments there.

The United States accepted a high degree of integration of its military units in the NATO military structure. Its units came under the command of non-US officers, as in the case of NATO Central Command, led by a German general. In addition US forces provided many communication and logistical assets necessary for the alliance military structure to operate effectively. (It can be argued that the US could easily accept the integration of its forces into the NATO military structure because US officers held the highest command positions, Supreme Allied Commander Europe (SACEUR) and Supreme Allied Command Atlantic (SACLANT).)

The US accepted the commitment, deployment and integration of a major part of its military forces in Europe largely because it could rely on allied participation in their primary mission, the defence of Western Europe. Moreover, the relatively large US force structure worldwide allowed it to consider (correctly or not) fighting simultaneously in more than one theatre, notably Europe and North-east Asia. Because the US could not count on support from European militaries in other theatres if war was imminent or already raging in Western Europe, it maintained forces in other theatres and at home for non-European operations. (Even during Vietnam, the US presence in Europe never fell below 200,000 troops.)[1] Thus the integration of US forces with those of its European allies would not significantly hinder US operational capabilities outside Europe.

US strategic thinking links the ability to lead a multilateral coalition with the ability to act alone if necessary.[2] Although unilateral capabilities may be necessary for multilateral coalitions, equally important is political cohesion among states. The habits of cooperation formed through military ties contributed greatly to the political cohesion of the alliance. Therefore, US leaders relied on the military for two contributions to the multilateral coalition with Europe: assistance in building the coalition before the alliance was called to act; and the ability to act alone to lead the coalition if necessary.

In the post-Cold War period protecting the United States' ability to act unilaterally around the world may require the withdrawal from Europe of some or all of the military deployments essential for an effective multilateral coalition with Europe. The US may not have enough forces to maintain a multilateral commitment in Europe separate from a commitment to unilateral action, if necessary, around the world.

Following the basic premise of the Regional Strategy, Secretary Aspin in the 'Bottom-Up Review' warned: 'History shows that we frequently fail to anticipate the location and timing of aggression, even large-scale attacks against our interests'.[3] He proposed that the US address these conflicts with a smaller force structure. By 1999, it should have 'sufficient forces to fight and win two nearly simultaneous major regional conflicts'. Such regional conflicts could include an Iraqi attack against Kuwait and Saudi Arabia and one by North Korea against South Korea. Aspin's goal of fighting two 'nearly simultaneous' conflicts contrasted with the Bush administration's goal of maintaining the ability to fight two major regional conflicts simultaneously.[4]

The US military presence in Europe must be increasingly justified by the rationales of maintaining stability in Western Europe and building coalitions for non-Western European security now that the immediate threat to Western Europe is gone. Although the Regional Strategy and the 'Bottom-Up Review' emphasise the need for a forward military presence and cooperation with allies whenever possible, the fundamental political requirement to protect US interests remains: multilateralism when possible; unilateralism when necessary.[5]

To comply with this, the US military must determine how to structure and deploy its assets. While the outcome of this assessment is important for all three of the military services, the US presence in Europe has historically consisted mostly of Army and Air Force personnel. In 1990, 68% of US military forces in Europe were Army, 28% were Air Force and only 4% were Navy. The permanent, shore-based US naval presence in Western Europe has always been small (about 15,000 ashore in 1990 compared with 18,000 afloat).[6] The Navy was essential to the defence of Europe, but played a less visible part in 'denationalising' Western Europe's military forces than did the Army and Air Force because of the preponderance of the US Navy within NATO and because the Navy's primary mission was off-shore. The most important Air Force mission was the delivery of nuclear weapons. This remains important, but no longer has the same political salience as the presence of ground and conventional air forces. The assumption that air forces could be deployed quickly to Europe from the United States has reduced the political significance of basing air assets in Europe. Because of their political importance, the following discussion gives greater weight to ground forces and their capabilities.

**Reasons for a continued military presence in Europe**
The United States is reducing the size of its military, but remains committed to defending its interests worldwide. In these circumstances, it should be looking to gather military assistance from as many of its allied and friendly nations as possible rather than emphasising unilateral action. There are several benefits for the United States of deploying a small fraction of its total military personnel in Europe. The US would improve the capabilities of the Western European states most likely to operate with its forces, thus allowing it to make better use of non-US military forces. The US is also more likely to have access to European facilities for non-European operations if it maintains military deployments in Europe.

MAINTENANCE OF ALLIED MILITARY CAPABILITIES
There was resistance during the Cold War, from both sides of the Atlantic, to the idea that Europe should gradually develop the capability to deter and defend against the Soviet Union with minimal or no assistance from the US. If Europe could defend itself, the transatlantic link would be unnecessary and the nuclear tie that was the basis of the doctrine of extended deterrence would come undone. In practice, how-

ever, the United States wanted to lessen its burden for Europe's defence rather than eliminate it completely.

The role of the US in maintaining European military capabilities was most obvious from the late 1970s. After the Vietnam War, the United States began to reassess the quality of American and allied military capabilities in Europe. The Carter administration launched the Long Term Defence Plan within NATO to upgrade NATO capabilities and raise each nation's spending on defence by 3% a year.[7] The Reagan administration adhered to these goals in the early 1980s, and there was a substantial improvement in the quality of allied armed forces by the end of the Cold War in 1989–91. US domestic pressures for a withdrawal of its forces from Europe would have been even stronger if improvements in European military capabilities had not taken place. US support for deployments in Europe would have been difficult to sustain if the allies were content to have less capable forces than the US and were not committed to assuming a greater share of the burden for defending themselves.

The quality and size of European forces is important to US security if its forces continue to be reduced. The allies of the United States in Europe are its most likely allies in military operations outside Europe. The European states also have some of the most modern and capable military organisations in the world. If the United States decides to emphasise multilateral actions, a US military presence in Europe will help ensure that the allies have forces able to assist it and help build multilateral coalitions through habits of cooperation.

INTEROPERABILITY WITH EUROPEAN FORCES

The premium on operating smoothly and efficiently with allied military forces, especially ground forces, will increase as US military budgets and force structures become smaller. Principal Deputy Under-Secretary Slocombe made the US view clear to the Senate on this point. 'Only forces that train together and serve together, share similar doctrines and procedures . . . and are practised in communicating and operating with each other – can have full confidence in each other and fight effectively together . . . we [the NATO members] also must work more closely together to ensure that our necessary reductions will not impair our national – and therefore our collective – capability'.[8] A permanent US military presence able to integrate well with other European forces will help the United States make the best use of European contributions in non-European operations.

If the capabilities and training of allied military forces diverge significantly from those of US forces, interoperability will become increasing difficult and European contributions to joint military operations will be much less effective. With fewer US forces in Europe, European states will be less likely to lend effective assistance to US actions. Thus a decision not to base forces in Europe will, over time, lead to an argument for why the US should not base forces in Europe.

During the Cold War, when most European allies were able to field a corps, there was no need for national militaries to work together below this level. A corps has sufficient combat support and combat service support units to enable it to fight relatively autonomously from other corps. Thus national militaries during the Cold War had to cooperate only with forces that were relatively independent of each other operationally. A notable, although small-scale, exception to this was NATO's Allied Command Europe (ACE) Mobile Force.

As the force structures of nations become smaller, some will no longer be able to field a corps on their own and will need to develop multinational units. Divisions within a corps are much more dependent on each other and on corps support forces, such as combat aviation and artillery, for their military effectiveness. Thus the requirement for greater real interoperability between national militaries will increase. Where two corps might have had to cooperate, two divisions within a corps have to do more: they must interoperate. For example, each of NATO's new multinational corps is led by a nation with responsibility for providing the command elements and support structure – artillery, medical and others – for the corps. A Dutch division in a German multinational corps must rely on German units for appropriate artillery support and medical care. Where two brigades within a corps are concerned, the reliance upon each other increases and thus the necessity for interoperability increases.

The cases of Belgium and the Netherlands demonstrate this trend well. During the Cold War, both of these states fielded a corps within NATO's layer-cake defence structure. Both states have since decided to cut their ground forces and will be unable to lead one of the new multinational corps in NATO's new defence structure. Instead they are relegated to providing forces to corps led by other states.

For the United States to operate with its allies, either in Western Europe or elsewhere, it will have to increase its ability to make use of military contributions below the corps level. In the Persian Gulf, even the largest of the West European allies, Britain and France, were unable to field a corps; they sent only divisions. Fortunately, the long history of training with British forces in Central Europe enabled US forces in the Gulf to use the British division effectively. A permanent US presence in Western Europe, large enough to be well integrated with several other allied forces into multinational corps, will allow the US to make the best use of allied military contributions in future contingencies.

## ACCESS TO EUROPEAN FACILITIES

Facilities in Western Europe have been critical to US military operations worldwide. US military planners assume that access to allied facilities in a crisis, difficult in any case, would be much less likely if it did not maintain a peacetime military deployment in Europe.[9] Moreover, the deployment of US forces in a crisis to locations where they did not operate in peacetime could reveal US intentions.

In recent years, the US has made good use of its facilities in Western Europe to support operations in the Middle East and Persian Gulf. During the bombing operation against Libya in 1986, the United States was able to use airbases and US FB-111 aircraft located in the UK which eased the need for carrier-based aircraft from the Mediterranean.

During *Operation Desert Storm,* a large proportion of US equipment deployed to the Gulf came from Western Europe.[10] Because Europe is an ocean closer to the Gulf than is the US, the travel time to the Gulf from Europe was less than the transit time from the continental United States.[11] The ability to stop over in Europe also substantially assisted the flow of equipment and supplies from the US to the Gulf. If casualties had occurred in substantial numbers, US military hospitals and allied civilian facilities in Western Europe were well situated to treat troops needing greater care than could be offered in a field hospital.

In future, when the US considers the wide range of contingencies that could occur without notice around the world, the ability to use bases in Europe would be important. Whether or not the US would gain approval to use its military facilities in Western Europe for military operations outside the region can only be answered in practice and not in theory. A peacetime deployment in a state, however, should substantially increase the chances of the host nation agreeing to the US request.

**Why end or reduce US military deployments in Europe?**

The United States is reducing the size of its military, but remains committed to defending its interests worldwide. Under these circumstances, it will need to maximise the ability of its remaining military forces to protect its interests. Assistance from allies and friends has been valuable in the past and is likely to prove valuable in the future.

The United States, however, cannot rely on the assistance of others to defend its national interests. Nor is it able any longer to devote substantial military forces to Europe in the hope that the Europeans may join US forces in operations beyond Europe.

## AVOIDING OPERATIONAL DEPENDENCY

Tighter integration of US and allied forces is a priority if defence budgets decline at a time when the most important contingency is the collective defence of Western Europe. Provided that all the members of a coalition can be counted upon to act, an integrated military structure makes the best use of dwindling military assets. As the most likely use of force is intervention in which states choose whether or not to participate, the participation of an ally will always be in doubt. Consequently, national militaries and their political leaders will not want to jeopardise their own ability to act by being dependent upon the willingness of other allies to act as well.

Although the US has been a strong advocate of NATO military integration and the division of roles among national militaries, it has also experienced some side-effects of these measures. The problem of German troops flying NATO AWACS aircraft to enforce the no-fly zone over Bosnia, discussed in Chapter II, is one result of creating a tightly integrated multinational force. In another measure to create a more rational division of labour among the allies in NATO, the US took on the mission of keeping open the Atlantic lines of communication, and other allies, notably Belgium and the Netherlands, assumed the coastal defence and mine-clearing mission. Because US defence planning focused on the defence of Europe, the US Navy gave much less emphasis to the mine-clearing mission.

When the US began to escort Kuwaiti oil tankers through the Persian Gulf during the Iran–Iraq War, it found itself desperately short of mine-clearing ships. NATO allies, whose contribution to the common defence of Western Europe was mine-warfare ships, were reluctant to send their ships to help clear these mines. The US was forced to activate old wooden-hulled reserve vessels which had not been used for decades. As a result, the US is now building new minesweepers.[12] Because governments would not use their own mine-warfare ships, the effectiveness of the operation was severely compromised. Although these multilateral arrangements are made with good intentions, participating states can withhold contributions and block, or disrupt, military actions decided upon by the other states.

## FORCE-PACKAGE FLEXIBILITY

Maintaining flexibility when assembling force packages is extremely important for the disposition of US forces in Europe. Military planning for contingencies assumes the ability to choose units and capabilities from the full inventory of the military structure. The contingency plans of the five US commands (EUCOM, CENTCOM, PACCOM, SOUTHCOM, LANTCOM) require various combinations of general-purpose forces (mechanised infantry, tactical fighters, carrier task forces) and units with more specialised capabilities (special operations units, combat aviation units, electronic warfare aircraft, *Aegis* cruisers). The combined requirements of these five unified commands exceed the total inventory for US forces – only acceptable because the US is not expected to fight in all five theatres simultaneously.

Each command is assigned a certain package of forces in peacetime based upon its most likely requirements in combat. High-value assets, such as combat aviation units or AWACS aircraft, may be useful in many different types of operations in different theatres, but are available only in limited numbers within the overall force structure because of their relatively high procurement, operational and maintenance costs.[13] Typically, there are many more general-purpose forces available for distribution than there are specialised forces available to fulfill the contingency requirements. Since the specialised assets are usually in extremely high demand, this allocation problem is a serious one.

Planning during the Cold War for a known, fixed, quantifiable threat against Western Europe made the allocation of these specialised assets easier. EUCOM, and to an extent LANTCOM, had first access to any systems entering the inventory and to the basing of high-value, specialised units. Large defence budgets also allowed greater numbers of these systems to be bought and maintained in the inventory.

Allocating high-value units in the post-Cold War period will be much more difficult. The Regional Strategy directs the US military to prepare equally for operations in many different theatres and for many different levels of intensity. No one region has priority. As the defence budget becomes smaller, the assets available will shrink and a greater proportion of units will be assigned multiple missions. This phenomenon will occur throughout the force structure, but will be most troublesome for those units in greatest demand and smallest supply. Specialised units, already dual- or triple-assigned, will receive more overlapping assignments.

Military planners will respond to these pressures for allocating scarce resources by maintaining the flexibility of the force. More specifically, planners will base fewer assets overseas. Instead, forces will be centralised as much as possible in the US where they can be given missions and redeployed quickly with as little difficulty as possible. The more useful the military asset and the scarcer the number of units available, the more likely it will be that the unit will be based in the United States.

Force flexibility can affect any type of unit in scarce supply. As the US force structure becomes smaller, more and more units will become 'scarce' or assigned to a large number of missions in different parts of the globe. In the extreme, even general-purpose units, such as armoured divisions and tactical fighter wings, will be affected and rationed.

The US Army is responding to these pressures by restructuring itself as a 'power-projection Army'. Previously, it was designed around forward-deployed units that were reinforced as necessary. In the 1980s, as much as one-third of the Army was based overseas. In future, the Army will have a greater proportion of its units based at home and will rely on its ability to deploy combat and combat-service support units quickly wherever needed. According to General Meade, Director of Plans and Policy for the Army, 'what you're seeing take shape now is an Army that is no longer forward-deployed, but rather stationed in the United States, including Alaska and Hawaii, and ready to project forces out'.[14]

The new model for the Army will be Central Command, a small headquarters in the United States that relies on US-based forces for rapid deployment overseas as it did during *Desert Storm.* These concepts are echoed in the latest version of the *Army Field Manual of Operations,* FM 100-5: 'Army commanders should assume that they will have to travel to the theatre and bring along all they need to operate in any environment'.[15] In Europe, it is possible that the Army will seek to maintain a forward presence without relying on forward-deployed forces.[16] Although there has been no official word from the Army, a forward presence might be maintained by sending units from the US to Europe for joint training with allies, or by rotating units from the US for short tours.[17]

The US Air Force articulated its vision for the future in 1991 in a White Paper entitled 'Global Reach, Global Power'. Like the Army, the Air Force plans to rely on US bases to project power and forces around the world. In Europe, the Air Force plans to maintain a forward-deployed presence of around 35,000 of the 100,000 US troops. According to General Boles, Director of Air Force Personnel, the Air Force is not as committed to Europe as it was in the past: 'when we bring a wing out of Europe, you're talking 72 aircraft and maybe 1,000 maintenance and support personnel that could be airlifted back pretty easy. And the air they have to fight in is pretty much the same over the United States as it is over Germany'.[18] Clearly, forward-based forces are not seen by the Air Force as essential to maintaining a forward capacity.

Basing forces in Europe will become more and more difficult if it limits their flexibility for global deployment. Forces assigned to EUCOM may be useful in Europe. However, deployments from Europe to the Middle East may require the consent of the host government in Europe and may be difficult to accomplish at short notice. Deployment of forces from Europe to the Pacific or Latin America would be even more complicated because airlift and sealift assets are likely to be based in the US. In the extreme, this trend could result in no US forces being permanently stationed in Europe, thus allowing them to be moved in the shortest amount of time to areas of conflict anywhere in the world.

POLITICAL FLEXIBILITY

For US military planners to be able to assume that units deployed outside the United States will be available for its own use is crucial. There is always a certain tension about the extent to which US forces overseas are there for the defence of the host country or for US operations elsewhere. During the Cold War, Germany was especially careful to describe the mission of the US forces there as simply the defence of Germany. The deployment of large numbers of US units from Germany to the Persian Gulf War helped to break this taboo. Yet US leaders must still plan for host governments blocking the US from using its forces based in their countries for missions other than national defence. When the US maintained a larger force structure, it could work around the restrictions of certain host governments. A shrinking force structure will make it more difficult to do so, should such restrictions occur. Unless host states guarantee unequivocal availability of US forces, there is much less incentive for the US to base its units overseas.

TRAINING IN-THEATRE

The need to maintain high levels of training for US armed forces has not changed with the end of the Cold War, but the need for and ability to undertake this training in Europe is in doubt. The need for training may thus become an argument for basing troops in the US rather than in Europe.

47

The operational requirements for forces during the Cold War put a premium on extensive training in the European theatre. The political requirements of defending German territory and involving allies in the initial defence resulted in US and allied forces being deployed rather thinly across a 900km front. The need to defend this front line with only 72 hours to 30 days of warning meant that an exceptionally high level of readiness had to be maintained.

Knowledge of the territory and extensive training were two of the most important assets for US soldiers in Europe. Ground exercises took place in training areas and across farm land. For the air forces, the most challenging mission was to fly low and penetrate the thick air-defence network in the east on the way to attacking targets. This mission required constant training at low altitudes over the West German countryside. For both ground and air forces, allied contributions needed to be integrated into a fighting force that could make best use of widely different national contributions and capabilities.

These ground and air exercises caused repeated discomfort for the civilian population, and open field exercises caused extensive damage. Although public criticism resulted in restrictions on training, training continued as long as the Cold War threat from the East was clear and the military advantage from training in-theatre obvious.

Allied forces in Europe are now no longer locked into a forward-defence position awaiting a short-notice assault by an overwhelming enemy just a few kilometres to the east. The likelihood of allied forces fighting a defensive battle in Germany is low; therefore, there is no premium on intimate knowledge of the countryside.

As the German people and political leadership are now much less willing to accept large-scale field exercises or low-level flying, the military is planning to restrict ground exercises to specific training areas, such as Hohenfels and Grafenwöhr. In order to lessen the noise for the surrounding communities and allow for greater manoeuvre training, German military planners also intend to limit the amount of live-fire training in these areas.[19] These restrictions are in addition to the common practice of not training at night. It is not yet clear whether they will apply only to German forces or, as is likely, to all military forces based in Germany.

Excessive restrictions on military forces in Germany will undermine the German commitment to multinational institutions and, in particular, multinational force structures. Multinational forces by definition require foreign participation. For Secretary Aspin and the Joint Chiefs of Staff, maintaining the readiness of US armed forces is an unbreakable rule in this time of declining defence budgets.[20] Restrictions on the ability to maintain the readiness of forces deployed overseas would be a powerful deterrent to such deployment.

Restrictions on training undermine arguments in favour of a forward military presence in Europe. Unless US forces are allowed to maintain their readiness while in Europe, they will be much less effective than their home-based counterparts for missions in adjacent areas, such as Libya. Unless US and allied forces in Germany are able to train, both alone and together, they will be much less effective in coalition arrangements. If they are unable to train adequately, US forces in Europe, regardless of their size and composition – 314,000 or 50,000 – will serve only as political symbols. It is not in Europe's interest to force the US to choose between maintaining a force able to respond quickly and effectively to contingencies all over the world, and deployments in Europe.

## SPEED OF DEPLOYMENT

Europe is much closer as the crow flies than the continental United States to North Africa, the Middle East and the Persian Gulf, regions where the US might need to deploy military force, thus proximity and speed of deployment should be strong arguments for basing US military assets in Europe. There are, however, issues that call into question basing forces in Europe for use in nearby regions.

Once political decisions are made to commit forces, great pressure is put on the military to move its forces to the area as soon as possible. There is an undeniable premium on speed of deployment, thus the time saved in deploying forces from Europe instead of the US needs to be looked at carefully.

Airlift and sealift have both strengths and weaknesses that make them more or less suitable to a particular mission. Airlift is fast and flexible and aircraft are able to deliver their payloads anywhere with a suitable runway. Its weaknesses are the small size of each payload and its high cost. The Air Force's largest transport plane, the C-5, has a maximum payload of only between 70 and 130 tons (depending on fuel carried).[21] The newest cargo plane, the C-17, is estimated to cost $350m each.[22]

The strengths of sealift are a huge payload and low cost. An SL-7 fast sealift ship can carry over 15,000 tons, approximately 120 times as much as a C-5, without costing 120 times the price (the SL-7 was bought on the used-ship market). A C-5 carries only one M-1A1 tank at a time, while an SL-7 carries the full equipment set for an armoured division.[23] The weaknesses of sealift are its slow speed and its ability to deliver cargo only to littoral areas.

As a general rule, airlift is best for moving small, light cargo quickly, while sealift is best for moving large, heavy cargo slowly. During the Persian Gulf War, airlift transported only 5% of the cargo, but 99% of the personnel. Sealift, by contrast, moved 95% of the cargo and only a small fraction of the personnel.[24]

In moving a brigade or division based in Europe or the US to the Persian Gulf, the apparent speed advantage of airlift is offset to a great degree by the high cost and small cargo payload of each aircraft. Because there are only a limited number of aircraft available in the US inventory (119 C-5, 265 C-141, 566 C-130),[25] and each aircraft can only carry so much per flight, flying a heavy division (armoured or mechanised infantry) from the US to the Persian Gulf would take over 21 days. Moving a similar unit by sea in fast sealift ships would take between 23 and 25 days. The number of round trips necessary for each aircraft to move the division all but eliminates the time advantage of airlift over sealift. If the task were to move a smaller unit requiring fewer round trips, such as a brigade, sealift would still take 23 days while airlift would need only 7–8 days.[26] Whichever is used, it will take between 21 and 25 days to move a large, heavy unit from the US to the Persian Gulf.[27]

Airlift of units from Europe to the Persian Gulf is faster than from the US, but only significantly so for heavy units least likely to be moved by air. Airlift of a brigade from Europe to the Gulf would take about four days instead of the 7–8 days needed to move the same unit from the United States. This 50% cut in time obscures the rather small improvement in absolute terms – four days saved out of eight. By contrast, airlift of a heavy division from Europe to the Gulf should take about 12–13 days compared to over 21 days from the United States. One week is an improvement that might have real implications in crisis-management situations.[28] However, as noted above, the vast majority of the equipment transported to the Gulf went by ship, including that from Europe. It is unlikely that a heavy division, rather than a smaller unit such as a brigade, will be moved by air from Europe. Therefore, the time saved in airlift for moving units from Europe as opposed to from the US, is most significant for those units least likely to be moved by air. For smaller units such as a brigade, more suitable as the initial response to a crisis, the nearness of Europe provides only marginal advantages in transit time.

Sealift from Europe to the Gulf is not, however, significantly faster than from the US. Unlike airlift, sealift does not travel as the crow flies. The apparent proximity of the Gulf to Europe as opposed to the US does not take into account the sea route necessary to transport forces in Europe to the Gulf. Sealift of equipment from Europe to the Persian Gulf would take about 20 days, compared with about 23–25 days from the United States.[29]

The reason for the similarity in sealift times is the location of US forces in Europe and the location of the sealift assets at the beginning of the operation. US equipment in Europe is predominately in Germany, not Italy, and the sea routes to the Persian Gulf from Europe and North America are the same once they pass the Straits of Gibraltar and enter the Mediterranean Sea. The difference, therefore, is in the time necessary to travel from Norfolk, Virginia, to Gibraltar as opposed to from Northern Europe, Antwerp or Bremerhaven. To deploy forces from Europe, sealift must either be hired locally or come from the US. If the ships come from the US, the deployment time from Europe could be longer than that from the US. The shorter time mentioned above (20 days) is for equipment moved in locally hired ships. The added advantage in time for sealift from Northern Europe instead of the US is relatively small – a few days – out of three weeks.

The United States and its allies could implement changes to increase the attraction of basing US forces in Europe for use elsewhere to make these forces more readily useable in other regions. Units stationed in Germany could be moved to Italy or Greece to be closer to the Middle East and the Persian Gulf, substantially cutting the deployment time by sea. It was partially with this in mind that the US moved a set of equipment to outfit an armoured brigade from Germany to Northern Italy. This brigade-set of equipment is available for deployment at short notice to all parts of NATO's southern region.[30] This, however, is only equipment and not personnel. Moving permanently stationed troops to Italy would be at the expense of those forces who assure the stability and security of central Western Europe, especially Germany. If carried out on a large scale, expensive facilities would be needed to house and train even half the 100,000 troops the US is planning to base in Europe. Nor is it certain that the Italian or Greek governments would welcome a huge increase in US forces in their countries.

In theory, the United States could preposition fast sealift assets in Europe in lieu of moving its forces from Germany to southern Europe. It is unlikely, however, that new ships dedicated to Europe will be purchased in the near future given the current budget climate. Reallocation of existing sealift assets from the US to Europe makes no sense when the bulk of US forces for deployment overseas is in the United States. Tailoring US forces in Europe for deployment to the Persian Gulf may slightly improve their transit time, but this is unlikely when the US is assigning multiple missions to its shrinking forces and endeavouring to make them as flexible as possible.

## Implications

Difficult political decisions must be taken by US leaders on multilateral cooperation with the European allies if its military presence in Europe is to continue. The US military structure may no longer be large enough to ensure unilateral capabilities while also devoting resources for possible multilateral actions with its allies. Thus operational considerations no longer strongly support maintaining US forces in Europe, as they did during the Cold War.

US foreign and security policy require two missions of the US military. First, it should support political and defence cooperation with the allies when possible. Forces should be stationed in Europe and integrated with those of other allies to protect Western Europe from internal instabilities and facilitate the building of transatlantic coalitions for non-Western European security issues. US forces help to prevent the renationalisation of defence and maintain habits of cooperation with allied military structures. Peacetime deployments in Europe make it more likely that the US will be given access to military facilities for actions outside Western Europe.

The US military presence in Europe must include a significant combat component to support multilateral cooperation with the allies when addressing sources of insecurity within Western Europe. These insecurities are related to fears about the renationalisation of defence arising from a breakdown of the Cold War integrated multinational defence structure of NATO. For the US to help maintain this structure, it must be able to participate throughout the multilateral force and with several other states in multinational combat units. Because the force must match up with many other allies, it must be large enough to do so. The alternative – deploying non-combat support forces or forces to receive US reinforcements – does not meet this criteria. The US would find it hard to form and lead a multilateral operation beyond Western European territory if peacetime cooperation had taken place only between non-combat service support units such as hospitals and kitchens.

The second mission of the US military is to retain the ability to act unilaterally to defend its interests. US unilateral action should not become constrained by or dependent on political decisions taken by allied governments. Thus US military planners should be wary of basing forces overseas or integrating with other allies to the extent that its forces are dependent on an ally for their effective employment.

There are two different approaches to reconciling these missions. It could be argued that the United States will not need to act alone provided it maintains multilateral political and military cohesion with its allies. Withholding a commitment to multilateral diplomacy because the United States believes it will not receive support from the allies is a self-fulfilling prophesy. Alternatively, it could be argued that the ability of the US to act alone allows it to hold together a multilateral alliance. However, maintaining the ability to act unilaterally should be its primary mission.

These two missions were easier to reconcile when the US force structure was large and could offset the possible recalcitrance of an ally. Yet as US forces are now becoming much smaller, the two political missions are no longer easily reconciled. Military planners are following guidance from the US political leadership and making their first priority the ability to act unilaterally.

A reduction or withdrawal of US military forces from Europe may be the unintended consequence of US policy priorities and a shrinking defence budget. If such a reduction is unacceptable to US leaders because of its consequences for US–European relations, US policies or budget levels should be reconsidered. Larger defence budgets are unlikely for the foreseeable future. At the political level, two issues might help solve this dilemma: whether the US has the capability to take unilateral action without at least passive support from the allies (allowing the US access to Western European facilities, especially air and naval bases); and, more important, whether the US could afford politically to take unilateral action without allied support.

The ability of the United States to undertake unilateral actions currently or in the future is a matter of debate. A smaller force structure will make operations on the scale of *Desert Storm* more difficult for it to undertake alone, even if it could have done so in 1990–91. Given sufficient political will and domestic support, reserves can be mobilised and actions taken. But the likelihood of acting alone will decrease as its cost increases. If support from allies can lower the cost of US action the likelihood of action should increase, and the best way to encourage allied support is to maintain a peacetime military presence in Europe.

For the US to take action without first securing the diplomatic support of its primary allies in Western Europe is a political decision. Without the overwhelming, omnipresent Soviet threat, the US political leadership may find it more difficult to convince domestic and world opinion of its need to take unilateral diplomatic and military actions. Recent US operations in the Persian Gulf and Somalia have been undertaken with allied support under the legitimising umbrella of the United Nations. UN approval has been cited by leading members of the US Congress as crucial for such operations in the future.[31] Even though there was some criticism of both the UN and the need to operate with other states in the calls for US withdrawal from Somalia, this does not mean that future US involvements will not need the sanction of the UN. The assumption in future may be that the US will not undertake unilateral actions. If so, then its military presence in Europe must be evaluated on the degree to which it encourages its allies to join a US-led multilateral initiative.

A change in US policy towards the transatlantic relationship could also reconcile the protection of its interests in Western Europe and other regions with a smaller force structure. If the United States saw the transatlantic relationship as an essential tool for addressing all security issues of common concern, a more multilateral approach to security policy would follow. Political guidance that assumed that a multilateral coalition was both politically and operationally necessary for the effective defence of US interests would cause a reassessment by US military planners. How to deploy forces would be evaluated for the political and operational benefits of supporting a multilateral alliance rather than for the operational considerations attendant on unilateral action. US leaders will need to evaluate the possible loss of its forces' military capability in Europe against the possible gain of access to allied facilities and assistance from allied militaries.

How the US deploys its military forces will be a key indicator of its broad-based acceptance of a new framework for thinking about US–European security relations and the role of the US military presence in Europe. A lasting military commitment will mean a firm political commitment by the United States to a more multilateral approach to its foreign and security policy.

## IV. DOMESTIC CONSIDERATIONS

Domestic considerations have always played an important part in US decisions about its forces in Europe. Planned and projected decreases in the US defence budget will continue to put tremendous pressure on all sectors of defence spending. In particular, the US Congress will scrutinise those items not in the United States on which money is spent. A complete or partial withdrawal of US forces from Western Europe is consequently not inevitable. To counter domestic pressures, Congress will need to be convinced that maintaining a US military presence in Europe is essential to address effectively its security concerns around the world.

### Defence budget and force-structure trends
US defence spending has been dropping in real terms since the 1986 fiscal year (FY).[1] For the foreseeable future, the important questions about defence spending are how much lower it will drop and how quickly. The most important consequence for the US military presence in Europe of a sustained drop in defence spending is the decrease in the size of the force structure.

The Bush administration projected that defence spending would shrink from about 5% of GDP during the Cold War to about 3% of GDP by FY95. The domestic agenda of the Clinton administration will require non-defence funding not planned for by Bush. Continued financing of an ever-accumulating budget deficit will also require substantial funding for which the defence budget is the most likely source. In the first year of his new administration, President Clinton has sought to shift money from the defence budget to other programmes. Secretary Aspin has set a target for cutting the defence budget by $88bn over four years (FY94–FY97) above and beyond the projected savings already planned by the Bush administration. For FY94 alone, Aspin cut the budget by $10.8bn.[2]

Making decisions about the defence budget involves trade-offs between force size, force readiness and force equipment. Aspin and General Colin Powell have consistently warned against budget decisions that would create a large but hollow force structure. Instead of maintaining a large force structure ill-equipped and inadequately trained to maintain its readiness, their goal has been the creation of a smaller, well-equipped and highly trained force.[3] Consequently, there has been a drop in the overall troop strength of the US from 2.2m during the Cold War to no more than 1.4m by the end of FY1997. The force structure is expected to shrink dramatically.[4]

| | FY91 levels[5] | Bush plan base force[6] FY94–95 | Clinton proposed FY94 force | 'Bottom-Up Review' FY99 force |
|---|---|---|---|---|
| Troop levels | 2.2m | 1.6 | 1.6 | 1.4 |
| Army divisions (active/reserve) | 26 (16/10) | 20 (12/6/2) | 20 (12/8) | 15+ (10/5+) |
| Air Force TFW (active/reserve) | 34 (22/12) | 26 (15/11) | 24.3 (13.3/11) | 20 (13/7) |
| Navy ships (carriers) | 530 (15) | 450 (12) | 413 (12) | 346 (11) |

This sharp drop in force levels gives new urgency to whether or not the US should maintain a permanent military presence in Europe. This is not only a strategic question for US foreign policy, it is also a financial issue of US domestic policy. Arguments in the United States in favour of a military presence in Europe must make clear that the gain to the national interest outweighs any potential gain in revenue for a state or Congressional constituency.

**Reasons for engagement**
THE BURDEN OF RESPONSIBILITY
Clinton administration officials believe that the US has a responsibility as the world's only superpower to lead the world. As Secretary Christopher said:

> Our responsibility to lead is probably greater now than it has ever been, because . . . we are the world's sole superpower. The United States will take action unilaterally when necessary. In some instances we will proceed with our friends and allies around the world, and in those instances, of course, we will consult with them. But our leadership is undiminished, and we are determined that the United Sates will fulfil its responsibilities in the world.[7]

The maintenance of a substantial military presence abroad to support US foreign-policy goals is part of the responsibility of being a superpower. The willingness to use this military power, whether unilaterally or with other allies, partly conditions the United States's position as the only superpower. Two manifestations of its leadership are the presence of US forces and equipment throughout most of Western Europe (Belgium, Germany, Greece, Iceland, Italy, the Netherlands, Norway (equipment only), Portugal, Spain, Turkey, the United Kingdom), and US officers in the two top military positions in NATO, SACEUR and SACLANT. Although political leadership does not require or imply military leadership, military leadership does imply political leadership.

Nevertheless, global responsibility as a rationale for its foreign policy has been suspect in the United States since the end of the Vietnam War. Whether the Clinton administration can make this the rationale for its approach to difficult foreign-policy decisions is not clear. It is clear, however, that some sort of rationale is needed to support US foreign-policy choices that have immediate costs, but only long-term benefits, such as US forces in Europe.

INFLUENCE

Americans often derided the Soviet Union during the Cold War for having nothing to offer other states except raw military power. Nevertheless, Americans also link the presence of US forces in Europe to US influence in European decisions on a wide range of political and economic issues. As General Galvin, then SACEUR, said in Congressional testimony:

Our European forward presence protects America and American interests and provides a link that gives us the opportunity to influence important decisions that affect our common defense. It is in our best interest to remain involved and influential in European affairs. In order to do so, we need a competent, credible, and operationally significant force in Europe ready for multifaceted missions.[8]

With the decline in the importance of military security questions relative to trade questions in the transatlantic relationship, the ability of the US to exact leverage for a military presence in Europe is likely to diminish. Europe is not interested in a broader political or economic basis for US engagement in order to maintain its influence in Europe. The European preference is for a US contribution to its military security, a source of declining influence for the United States. Nevertheless, hope for a useful link between military and economic relations remains strong in the Congressional debate.

## DETERRENCE AND DEFENCE

The use of overseas deployment to deter future conflicts involving the US is strongly supported in the United States. Deterrence of aggression against Western Europe may remain a long-term strategic goal, but no longer seems an immediate problem. If deterrence is to be a credible reason for maintaining US troops in Europe, the threat to be deterred must also be credible. Either Russia still poses an immediate threat to Western Europe, or Western Europe is at risk from internal sources of instability. Neither threat is likely to have enough credibility in the United States for deterring conflict in Europe with a permanent US military presence.

### Pressures for disengagement

There are several sources of domestic pressure for the withdrawal of some or all US forces from Europe. Most of these result from the need to allocate a shrinking defence budget.

## THE ADDED COST FOR OVERSEAS BASING

The added cost of maintaining forces overseas makes these deployments a target for cuts when the defence budget as a whole is shrinking. Credible comparisons of the costs of basing equivalent forces in Europe and the US are difficult to make. Infantry battalions in Texas may be the same as infantry battalions in Germany, but the local costs of maintaining them may be quite different because of energy prices, transport costs and the local cost of living. The military mission of the two battalions is likely to be quite different, which also leads to different costs. Traditionally, US forces overseas have maintained a higher readiness level than forces based at home, and the greater requirements for training and maintenance of equipment contribute to the increasing operating cost of overseas forces.

During debate in the winter and spring of 1992 on the stationing of US forces in Europe, General Galvin, NATO's Supreme Allied Commander Europe and US European Commander, offered some broad figures for the added cost of basing US forces in Europe. (Galvin supported the deployment of US forces in Europe and therefore had an interest in presenting Congress with low figures for the cost of basing these forces there.) As he reported, the cost of stationing 150,000 troops in Europe rather than in the continental US is about $500–700m per year. This is the added cost, about 10% of the basic cost of the forces themselves, above and beyond burden-sharing payments and other contributions from host governments.[9.]

Burden-sharing costs are an extremely sensitive topic in both the United States and Europe, especially in Germany. The US has been moderately successful in negotiating various NATO and bilateral host-nation support agreements with European governments to cover some of the costs of stationing forces overseas. With Japan and Korea, the US has been outstandingly successful. By the mid-1990s, the Japanese and South Korean governments will provide payments to cover all added costs for the US forces they host. Indeed, basing US forces in Japan and South Korea may even be less expensive than maintaining the same forces in the continental United States.

Comparison of European burden-sharing and burden-sharing with Far Eastern allies is difficult. Western European governments provide a proportionally greater share of forces and capabilities to their alliance with the United States than do Far Eastern governments. Western European states also spend a much greater proportion of their GDP on defence than does Japan, and at least as much as South Korea.[10] Nevertheless, the success of US–Japanese and US–Korean negotiations has prompted Congress to pressure Europe for agreements to reduce further costs for stationing US forces there, agreements that Germany is not interested in making.

The total cost differential for basing in Europe, less than $1bn, is relatively modest when compared with the US defence budget of $261bn for FY94. At a time of overall decline in defence spending, however, all elements of the defence budget are eligible for cuts. Again, the burden of proof tends to be on those who advocate deployments overseas to justify their added cost, however modest.

## LOSS OF MILITARY PRESENCE IN THE US

Basing military units in the United States rather than abroad brings the economic benefits of military payrolls to Congressional constituents that vote. Members of Congress consistently call for the redeployment of military forces to the US and for keeping military bases there open. These two elements are intertwined: part of the rationale for redeploying forces is to keep the bases open; and keeping a base open requires the deployment of a military unit to the base in question.

Military bases and the units stationed there infuse relatively large sums of money into local economies through payrolls and spending on local services. As a result of the most recent series of base closures, California will lose over 78,000 military and civilian jobs at military installations. The total payroll lost by California is estimated to be about $4bn per year.[11] This money, not raised through local taxes, is perceived by local people as a net addition to the local economy, coming as it does from the Federal Budget. The social dislocations caused by abruptly cutting these funds can be enormous.[12]

Any attempt to close a military base in the US is highly political and fraught with difficulties. No Congressman or Senator wants the responsibility of eliminating bases in his or her own Congressional district. Nevertheless, the deep cuts that are being made in the defence budget and force structure make it hard to argue against closing some bases. To eliminate any political liability for voting to close bases, Congress created a special, non-partisan Base Closure Commission in 1990. Once the Commission recommends a list of bases to close, the Congress and President can either accept or reject the entire list, but not amend it.[13]

By comparison, Congress has made it easy for the Executive Branch to close an overseas base. Congressional approval is not required to close an overseas facility, unlike the complex procedures devised for closing bases at home. This asymmetry in the process for closing bases in the US and abroad reflects the general political pressure in the United States to close at least as many overseas facilities as at home. In the most recent announcement of closures, June 1993, the independent Commission recommended the closure of 129 bases: 92 (71%) overseas, of which 85 were in Europe. Since 1990, when the current closure system was put in place, the US has announced large reductions in staff or complete closure of 840 bases overseas to be implemented by 30 September 1996. Of these facilities targeted, 740, or 88%, are in Europe.[14]

This emphasis on disproportionate reductions overseas is also reflected in plans for the force structure. Under the Bush administration plan to cut the force structure by 25% over the years 1990–95, troop levels in Europe would drop by slightly over 50% (from 314,000 to 150,000), while troop levels overall would drop only 27% (from 2.2m to 1.6m). In 1992 Congress mandated a ceiling on US troops in Europe of 100,000, making the cut there a huge 68%.

The result of these measures for the US presence in Europe could be devastating. Congress's goal is to lessen as much as possible the effect in the US of reductions in its force structure, by closing a proportionally larger number of overseas bases and increasing the proportion of the force structure based at home. Financial reasons for basing fewer US forces overseas were best summed up by the anonymous Congressman who said, 'Remember, the Europeans don't vote'.

## BOSNIA AND THE LOSS OF US LEADERSHIP

Congressional frustration with the Clinton administration and Western Europeans over lack of forceful action in Bosnia is threatening support both for NATO and for the US military presence in Europe. Some Senators and Congressmen question the maintenance of any US forces in Europe if the Europeans and the US cannot agree on whether to use them in places like Bosnia. Instead of multilateralism where the US follows the lead of its allies, some members of Congress think the US should lead the alliance as it did during the Cold War and more recently in the Persian Gulf and Somalia.

The Clinton administration argues that Bosnia is a European problem. The US will act there only in concert with Europe on a multilateral basis. The administration also maintained that the US would not put its ground forces at risk in Bosnia without an agreed peace plan.[15] This position has been strongly supported by Congress. Instead of a commitment of ground forces, the administration has advocated lifting the arms embargo and making air strikes against Serbian forces.[16] The Europeans have argued that air strikes would risk retaliation against their troops on the ground and would interrupt the flow of humanitarian aid. On these grounds, Europe has blocked US proposals by denying the unanimous consensus necessary for action in a multilateral alliance.[17] Unwilling to risk ground forces and unable to gain a consensus on air strikes, US policy on Bosnia came to a halt.

In an effort to deflect attacks on the administration's policy, Secretary Christopher has blamed the Western European allies for blocking US initiatives: 'We continue to regret the fact that the Europeans could not be persuaded' to support proposed US initiatives to lift the arms embargo and undertake air strikes in Bosnia.[18] This tactic, however, only served to focus criticism on Europe and on US support for European security. In a May 1993 letter to President Clinton, Senators Lugar and Dole sharply criticised European policies.

> In recent years our allies have argued for the maintenance of a significant number of American troops in Europe as a means of preserving European stability. However, the inability of NATO to act effectively to contain and stop a major war on European soil is bound to raise grave doubts among both the American people and the Congress about whether the enormous investment we make in NATO is reaping sufficient benefits.[19]

Congressman Barney Frank was more blunt: 'Bosnia has shown there is no real role for US troops in Europe'.[20] Because of Congressional unhappiness with European inaction and the administration's unwillingness to exert greater pressure on Europe for action in Bosnia, the US military presence in Europe is in danger.

## CONGRESSIONAL INITIATIVES

There have been two notable Congressional initiatives to address the concerns raised in this chapter about US domestic support for maintaining its troops in Europe. Congressman Frank sponsored a bill to increase substantially the amount of money European states would pay for the privilege of hosting US forces. He argued that the failure of the European allies and the United States to agree on a plan to resolve the Bosnia crisis shows that the only role for US troops in Western Europe is to defend the Western Europeans. In this case, the allies should pay all the costs of maintaining US forces for their own defence.[21] Under Frank's bill, the United States would for the first time request compensation for at least 50% of the personnel costs of its forces based in Europe. In addition, 75% of non-personnel costs – utilities, military construction, environmental clean-up – would have to be covered by the host state. The legislation, if enacted, would require the President to negotiate these terms with the European host states and would prevent US money covering these costs if the President failed to secure European agreement by FY96.[22]

Frank's bill goes beyond reimbursing the United States for the added cost of basing forces in Europe rather than at home. The Western European states would be required to pay a portion of the basic personnel costs of keeping these troops in uniform, a cost that the United States has to pay regardless of where its forces are stationed. Because the legislation requires the elimination of all funds for US forces in Europe if an agreement cannot be reached, this is a blunt attempt to pressure the Western European states into paying more for hosting US forces. As most US forces in Europe will continue to be based in Germany, the additional costs will be felt most heavily there. This is a high-risk strategy for the US to adopt at a time when Germany, and Europe in general, is deep in economic recession.

Senator Lugar has taken a different approach to restructuring the US–European security relationship in an attempt to gain domestic support for US military involvement in Europe. In speeches and articles, he has argued for reform of NATO to address the new strategic environment in Europe: 'The common denominator of all the new security problems in Europe is that they all lie beyond NATO's current borders'. To survive, Lugar argues, 'NATO will either develop the strategy and structure to go out of area' or it will 'go out of business'. However, Lugar's approach limits NATO's new horizons to Eastern

Europe and the territory of the former Soviet Union. While mentioning briefly an 'arc of crisis to the south' of Europe, Lugar focuses on the East. Having developed a strong argument for why the United States and Western Europe must work together beyond the old boundaries of NATO, Lugar does not expand the geographic bounds of this new transatlantic bargain beyond Europe, only beyond Western Europe.[23] It is also unclear what legislative mechanisms he might use to bring pressure to bear on the President and the Europeans other than that explicitly invoked by Frank in his bill: the withdrawal of US forces.

## IMPLICATIONS

The US armed forces will continue to decline as long as the defence budget does, and as long as the emphasis within the Department of Defense is on maintaining the readiness of US forces rather than on their overall size. The task facing the Clinton administration and Congress is to allocate shrinking pools of both troops and dollars. Congressional pressures to withdraw forces from Europe and close bases overseas in greater numbers than at home are quite strong.

In the 1970s, Congressional proposals sponsored by Senator Michael Mansfield to cut US forces in Europe could be countered by references to the Soviet threat to the United States and Western Europe. Now, however, there are fewer counters to domestic budgetary considerations. In future, the most convincing argument for a US presence in Europe will remain that it is essential for deterring aggression against, and defending if necessary, vital US national interests. A strong transatlantic relationship for common defence of common interests, wherever they are under threat, is the best way for the US to make deterrence credible. Unfortunately, because of US and European policies in Bosnia, a multilateral approach to security has become synonymous with lack of US leadership and troublesome European restraints on US proposals to act.

If the US follows its stated preference of defending its interests unilaterally, the only rationale for its forces in Europe is to defend its interests in Europe and adjacent areas. The necessity of basing US forces in Europe to operate in adjacent areas has been called into question by the military, leaving few persuasive arguments for keeping US forces in Europe in a time of declining spending and military capabilities.

The Congress, like any elected body, prefers continuity to dramatic change. Congressman Frank's proposal is a radical step that reflects a widespread and deep unease in the Congress over the domestic effects of defence spending cuts. These cuts will continue. If it is possible to counter these strong domestic pressures, there must be a clearer articulation of US security interests and the role that Western European allies could play in their defence. If it is not possible to counter domestic pressures to cut the defence budget and the US force structure becomes smaller, the rationale for seeking European assistance to defend shared interests becomes stronger. A crucial step in gaining Western European assistance will be the maintenance of a permanent US military presence in Europe.

# V. INSTITUTIONAL IMPLICATIONS

If NATO is to repeat its past success, reform must reflect the underlying changes that have taken place in the transatlantic security relationship and in European security challenges. NATO will remain relevant only as long as its members find it of use in addressing their vital national interests. This chapter asks whether NATO can be developed so that the transatlantic security relationship continues to address the fundamental security interests of its members. Only in this way will the relationship remain relevant. If the US continues to play a role in European security, what should that role be? What part does Western Europe have in the US security agenda? What part do US military forces in Europe play in keeping the transatlantic relationship politically credible and militarily useful?

## Why NATO?

In spite of an overlap in membership and similar charters, NATO and the WEU have different purposes and are not in direct competition. The WEU was founded as a response to post-war concerns about German re-armament. Since 1989, the WEU has been incorporated into the European unification framework and given a new purpose. It is the embodiment of the aspiration for a European Security and Defence Identity (ESDI) and thereby acts as the forum for the development and implementation of security and defence policy in the governmental structure of the European Union.[1] The relevance or irrelevance of European security and defence policies to the problems of European security will not affect the survival of the WEU. Rather than existing to respond to the changing security environment in Europe, the WEU is an essential part of the political process of European integration. As an organisation it may have lost some autonomy through its incorporation into the European Union, but it has gained permanence.

NATO, by comparison, derives its political importance in Europe from the participation of the United States. Without it, NATO has only a few more capabilities than the WEU, but has less political salience. The future of the US role in European security, therefore, is of crucial importance to the future of NATO as a viable and credible organisation. For reasons that will be discussed, NATO may not only be the best way of remaking the US role in European security, it may also be the best way of developing a European role in US security policy.

**Review of the issues**
US STRATEGIC CONSIDERATIONS
Of the two rationales for US engagement in European security –
addressing indigenous sources of insecurity in Western Europe (such
as fears of post-war German rearmament) and building coalitions for
non-Western European security – the former is a powerful interest, but
a difficult issue to make a high priority for US resources. It has become
increasingly difficult for US leaders to argue that the US must continue
to spend its defence budget on remaining engaged in Western Europe
lest devastating wars occur. Nor can such a position last long as the
primary rationale for an alliance of equals. The future role of the US in
European security depends upon acceptance by both the United States
and its European allies of their relationship as a partnership for addressing security outside Western Europe. This can mean a partnership
for addressing security in Eastern Europe, Russia, the Korean penin-
sula, the Persian Gulf and the Middle East. The alliance might address
these security problems differently, but they are all beyond the bounds
of Cold War NATO. As these states remain outside NATO's membership and beyond a collective defence commitment, coalitions of current members will be the rule for addressing these security issues.

The United States now needs to decide how to address its new
priorities among its interests, alone or with its European allies.
Multilateralism was useful and effective during the Cold War: an
external threat ensured a high degree of political unity among the allies
for NATO's primary mission, the defence of Western Europe. Transatlantic coalitions for anything other than territorial defence have been
more problematic and are likely to remain so. The greater the degree of
choice over taking part in a multilateral operation, the greater the
uncertainty that a state will participate. The end of the Cold War
allowed the Western European states to join a 'coalition of the willing'. But the inherent unpredictability of forming coalitions for actions
other than territorial defence is also a challenge for the United States.

If the US chooses unilateralism because coalitions are too unpredictable, NATO has little or no future as a viable and credible organisation. US engagement in Europe cannot be sustained for the purpose
of addressing highly unlikely sources of instability within Western
Europe. Without US engagement, NATO will become dysfunctional
and irrelevant to European security. There is no reason for Europeans
to use it as a forum for discussion and consultation on security matters
if the WEU can serve as well. The WEU has the additional political
advantage of being the forum of choice for staunch supporters of
European Union.

If the US responds to the challenge of forming coalitions with Europe for security issues outside Europe with a new emphasis on multilateral diplomacy and military action, the best way forward is to propose a new purpose for the US–European security relationship. NATO should then become an alliance bounded by shared interests rather than shared geography. The problem of ensuring allied participation in multilateral initiatives will never be settled to the same degree as it was during the Cold War. Through NATO, however, the US has the best opportunity to build a strong peacetime consensus on security issues and lessen the chances of a coalition member declining to participate. NATO is also the obvious organisation for the United States to use to continue the close operational integration between it and its European allies. It is also the best way for the Europeans to keep the US engaged substantively in their security.

## WESTERN EUROPEAN SECURITY POLICIES

The crucial Western European governments – Britain, France and Germany – think that the US role in European security was, is and should be closely tied to NATO. They are interested primarily in maintaining their Cold War security arrangements in which the United States had an essential role. The goals of Britain, France and Germany for the US in Europe are thus conservative in that they seek to preserve the status quo.

Western Europe wants the United States to continue to guarantee Europe's security from external and internal sources of instability. This goal explicitly includes the stationing of a substantial number of US combat troops in Europe, as these troops add credibility to the US commitment. They are also essential to the building of multinational units that embody the European goal of preventing renationalisation of defence in Europe.

NATO is the organisation through which the United States can best contribute to European security. It is also the organisation through which the Europeans can best circumscribe and control US influence in Europe. The Western Europeans do not want the US to have any more influence in Europe than necessary. Nevertheless, they are anxious to preserve its role in their security. A role in European security through the EC is not an option. Through NATO, the allies can solve this problem. In NATO, the United States has a place within a European institution that is not the EC, and makes the essential contribution to European security requested by the Western Europeans.

Missing from European thinking is a full understanding of why the US would accept the role in European security offered by the Europeans. Of the two rationales for US engagement in European security, Britain, France and Germany want a US role primarily to provide security in Western Europe. Western Europeans are much less interested in and less able to participate effectively in coalitions beyond their borders. Their ability to make implementable decisions even in the former Yugoslavia has been poor. According to their defence budgets, Britain, France and Germany will correct few of the deficiencies their forces showed in Yugoslavia and the Persian Gulf. In many cases, Western Europe will have less capable forces overall in the future. It is unlikely, however, that these will have to be employed in Western Europe. It is much more likely that the UN or the US will ask the Europeans to send forces to the East, South or Far East. For such missions, European capabilities have been limited and are becoming more so. These regions, however, are likely to become the regions of highest priority for US security and defence policy. If the Western Europeans remain uninterested in security beyond their continent, US interest in working with Western Europe will wane to the detriment of European security.

OPERATIONAL CONSIDERATIONS

The degree of importance the US military gives to NATO, and by association to allied militaries, corresponds directly to its political guidance, the missions assigned it and the forces available. US political guidance during the Cold War and into the present has directed the uniformed military services (Army, Navy, Air Force and Marines) to act with the allies if possible and alone if necessary. To support this dual approach the US military needed forces in Europe to maintain links with allied forces for coalition operations, and sufficient forces to act alone if the allies did not join in non-European contingencies. During the Cold War, there were sufficient forces in the US structure to cover both missions independently.

The missions for which the US military must now prepare have multiplied and are much less predictable. At the same time, there are about 36% fewer US soldiers available for these missions. Military planners are responding to these changing circumstances by reallocating forces to improve their deployment flexibility and maximise their effectiveness. Under these new circumstances evaluation of the costs and benefits of basing US forces in Europe yield equivocal support, at best, for continued deployments. In current plans, US force levels in Europe will drop by at least 68%, to no more than 100,000. They may even go lower.

The dramatic decrease in US military personnel overall could undercut US political guidance to the military. When forced by necessity to choose between the two parts of their political guidance, US military planners are structuring forces to ensure they can operate alone, rather than emphasising multilateral operations with allies. The ability of US forces to act with the Western Europeans will be much weaker without a permanent presence to maintain the day-to-day habits of cooperation that make a coalition force effective. In the long term, the lack of a US military presence will severely decrease the usefulness of NATO both to the Europeans and to the United States.

Whether the United States will ever undertake action without first securing the diplomatic support of its European allies is a political decision with consequences for NATO. However, the reallocation of US military resources is forcing a decision concerning the policy of multilateral diplomacy and military action. The current political guidance may remain unchanged. If the military is directed to maintain a specific portion of US forces in Europe, it can do so only at a cost to the overall operational effectiveness of its forces if they are directed to act alone.

Conversely, the US political leadership could change its guidance and place even greater emphasis on multilateralism. Some of the negative consequences of deploying forces in Europe would be mitigated by an increased likelihood of forming coalitions. If this is the chosen political path, it leads straight to NATO, and it is at NATO that the US has had its greatest success in developing consensus positions on security policies and in translating those policies into an effective, integrated multinational force structure.

## US DOMESTIC CONSIDERATIONS

NATO has both positive and negative associations for US Congressmen. Positively, it is associated with the exercise of US responsibilities as a superpower during the Cold War. As even some French officials have put it, US worldwide leadership is based upon its leadership in Western Europe,[2] and NATO is greatly associated with US engagement in Europe, specifically the deployment of US military forces to deter a third European war this century.

NATO's Cold War legacy also has negative associations in the United States. While 'defence of NATO allies' was a catch-phrase over many decades for the complex US role in Western Europe, 'US support for NATO' is now associated by some members of Congress with spending US resources to defend rich European allies from a virtually non-existent threat.

It may be impossible to overcome completely the strong fiscal pressures in the United States that are influencing its military policy. The best possible approach is to continue to rebalance the transatlantic security relationship to give equal attention to security concerns beyond Western Europe. If the argument for engagement can be won, building on the US role in NATO will be a central part of the plan. NATO's strategic mission will have to be visibly altered in changing the terms of debate about the US role in European security. NATO, as the institutional embodiment of this new transatlantic relationship, must be redirected to become the forum for a US–European partnership for worldwide security issues.

### NATO reform
PROGRESS TO DATE

At the London Summit in July 1990, the member-states announced their intention to 'enhance the political component' of the alliance and 'to build partnerships with all the nations of Europe'.[3] The changes, as they have continued to be made through the January 1994 summit, however, stop short of casting the US–European relationship as the key to addressing the post-Cold War world beyond Europe.

The London Summit Declaration set in motion a series of important political and military changes to NATO. A new Strategic Concept, adopted in November 1991, broadened its mission from defending its members to include dialogue and cooperation with its traditional adversaries, Eastern Europe and the Soviet Union.[4] The North Atlantic Cooperation Council (NACC) was also established in December 1991, and all former members of the Warsaw Pact and the Soviet Union were invited to join. NATO began a fundamental restructuring of its forces and command structure from which came the formation of multinational corps and a rapid-reaction corps.

Although important, these changes are not sufficient to reflect a new agenda for transatlantic security. Secretary Christopher, at the June 1993 meeting of NATO Foreign Ministers, raised five issues with his counterparts: maintaining NATO's military strength; improving its peacekeeping ability; ensuring it works with other organisations such as the UN and CSCE; using NATO to promote security across all of Europe, West and East; and using it to address threats beyond Europe.[5] The US agenda for NATO is not thus focused primarily on Western Europe, but on addressing security beyond the alliance's traditional territorial limits. The alliance has a role to play beyond the territorial defence of its borders. In Western Europe, this wider vision may extend only to a larger Europe. Washington's initiatives adopted at the January 1994 summit – Partnership for Peace (PFP), CJTF and greater allied efforts against the proliferation of weapons of mass destruction – reveal that Washington's vision for NATO extends well beyond Western Europe to Eastern Europe and off the continent completely.

## OPTIONS FOR FURTHER REFORM

Lasting reform of NATO requires the transatlantic relationship to address the security issues of greatest importance to Western Europe and the United States. If such reform is successful, the continued relevance of NATO and the US presence for European security is assured. Otherwise the alliance's relevance will remain in doubt, its credibility in deficit. There are two basic options for renewing the transatlantic relationship and keeping NATO relevant: using it to address security issues related to Eastern Europe; and using it to address security issues of common concern wherever they are threatened. These are not incompatible options. NATO approved US initiatives that might in time lead to both during the January 1994 summit.

Britain, France and Germany are not seeking changes in the US role in European security other than limited reform of NATO. The Western Europeans have articulated their view of the US role in European security: a US contribution to European security, of which an essential part is a permanent military presence, circumscribed by the narrow political bounds of NATO. The British, French and German leaders acknowledge that NATO and the US role in Europe are essential to European security. Thus, these states want them to be relatively static and provide an element of continuity. NATO currently addresses the most basic and important security concern of the West Europeans – territorial defence. That PFP with non-NATO European states was a US proposal only highlights that in Western Europe, the burden of proof is on those advocating a need for change in the transatlantic relationship and the US role in European security. Western Europe may accept incremental change, but only as the price of continuity in their security.

In spite of statements of support for NATO from the Executive Branch, among many members of Congress NATO's credibility deficit is deep. A new rationale for the US presence in Europe is needed. Western Europe is no longer the US interest under greatest threat. Therefore, there is a greater crisis of relevance for NATO in the United States than in Western Europe. Which of the two options for reform of NATO better addresses its credibility problem in the United States?

## NATO FOCUS ON EASTERN EUROPE

The United States and Western Europe could use NATO to address their security concerns related to Poland, the Czech Republic, Hungary and possibly Slovakia. These Eastern European states are worried about their security and want to join NATO. Their main concern is the threat from the recently departed Russian Army, although tension within their region, between NATO and Russia, is also a concern.

NATO could either allow the Eastern European states full membership, or greater participation in the practical aspects of membership. In 1991, the NATO allies put off the political question of membership for their former Warsaw Pact adversaries, and instead set up NACC through which some of the benefits of NATO membership could be extended piecemeal to non-members. The same problem confronts NATO in 1994.

In principle, it should offer full membership to Eastern European states only if the alliance's present members are willing to defend the territory of the new members. In other words, the present members will expand NATO only if they think it will enhance their security. Membership for Eastern European states may deter aggression within and against the region. New members in Eastern Europe would give the Western European states, and Germany in particular, the ability to defend themselves against Russia far beyond their eastern borders.

Alternatively, incorporating former Warsaw Pact states into NATO may make Russia much more difficult to deal with. Instead of lessening tension and the possibility of aggression, expanding NATO membership could well cause the reverse. (Offering Russia membership in NATO might solve this problem. Few NATO members, if any, however, think this would increase their security.) This leads back to the question of whether the present members of NATO are willing to defend Eastern Europe. Western Europeans are first and foremost concerned with their own territorial security. An increased possibility of conflict in Europe, even in Eastern Europe, is enough to deny membership for the Eastern European states. Only Germany might take this gamble because of its 'frontier' position, but it might be overruled by its allies.

Instead of membership, NATO might offer non-NATO European states a combination of limited political commitments and practical military measures. This would garner as many of the European security benefits of new members in NATO without a formal commitment to defend Eastern Europe.

NATO's new Partnership for Peace builds on the NACC concept by offering the *de facto* extension of Article IV of the North Atlantic Treaty to all states of Europe and by offering to increase the level of military cooperation non-members can pursue with NATO. All European states willing to participate in PFP have to file implementation plans explaining how they intend to contribute to partnership activities such as military exercises and planning, improving transparency in defence budgeting and greater civilian control of their defence establishment. To facilitate the ability of PFP states to join NATO military exercises or operations, they are asked to provide officers for a military planning cell. States would each pursue their own level of participation and therefore choose for themselves the extent of their engagement with NATO. For example, states will decide how often and with what size of force they might participate in exercises. NATO allies and states participating in PFP would agree to consult whenever the territorial integrity, political independence or security of a partner state was threatened, a *de facto* extension of Article IV.

Extending the application of Article IV of the North Atlantic Treaty to NACC is a new, limited political commitment for Eastern Europe. This has more symbolic than real importance because NACC already serves this function. Nevertheless, the link to the NATO Treaty would give PFP and NACC greater legitimacy.

NATO's offer of military measures to improve the ability of PFP states to defend themselves also enhances their ability to contribute forces to operations undertaken by NATO. They would also be improving their ability to make a net addition to collective defence if they become members.

In addition to the joint military exercises NATO might offer the Eastern Europeans, NATO should initiate a defence planning process with PFP states analogous to that run through the Defence Planning Committee that already exists among the 16 members of NATO. NATO could offer to evaluate defence budgets, force planning and procurement programmes on the same rigorous basis as the evaluation undertaken by the Defence Planning Committee. Critical evaluation by a committee of peers may be a harsh learning environment for the Eastern European democracies, but is likely to prove more effective than bilateral consultations behind closed doors. Although not targeted at the Eastern European states, PFP is explicitly designed to develop into a varied set of relationships depending on the extent of engagement a state chooses to pursue. It is assumed that the Eastern European states will choose to test the limits of their NATO partners and pursue as deep a level of engagement as possible.

## WOULD THIS SECURE NATO'S FUTURE?

What gain will there be for the credibility of NATO among its current members from PFP? Neither Britain nor France see a great need for NATO to focus its attention eastwards. Britain supports extending NATO eastwards only if this helps to keep the US focused on European security. France wants to ensure that NATO does not go where the EC and the WEU have not gone before. Germany is worried about developments in Eastern Europe and Russia and wants help from any and all parties, the EC, NATO and the CSCE. Germany is already a strong supporter of NATO and the US military presence in Europe. NATO and US involvement in Eastern Europe, provided it did not upset Russia, would be welcome in Germany, but is not necessary for NATO's credibility. The Western Europeans, however, are unlikely to want to increase their military expenditures to provide added security for Eastern Europe.

Eastern Europe is a security issue for the United States. An important victory of the Cold War was freedom for the states of Eastern Europe from Soviet domination. Losing the opportunity for democratic and economic reform in Eastern Europe would be a severe setback for US foreign policy. Eastern Europe is also the frontier of Western Europe and greater security in Eastern Europe thus provides greater security for Western Europe. This notwithstanding, the security of Eastern Europe alone should not be the major new focus of the United States' most important tool of security policy when its fundamental security interests are threatened as they are in the Persian Gulf, the Middle East, South Korea and Japan. The burden of protecting US interests in Eastern Europe can be shared to a greater degree by Western Europe. At the same time, however, in other parts of the world the US has fewer local allies able to help defend its interests and thus can benefit more from help given by its closest European allies.

NATO is in danger of irrelevance; using it to address the security of Eastern Europe helps, but does not address primary US national interests. PFP in its own right is an important step for NATO and a significant improvement in the security of Eastern Europe. However, the security of Eastern Europe relies upon US involvement in European security, whether PFP works or not. If US involvement is not assured, neither is Eastern European security. NATO, and US engagement in European security, may have a new lease of life, but only a short one at best.

74

## NATO AND COMMON INTERESTS

NATO must be remoulded into a transatlantic alliance to defend common security interests wherever they are threatened. This option uses the US–European security relationship to address all or most aspects of the new Regional Strategy. The US role in European security would become the linchpin of their partnership for global security. This policy would then require the European allies to take a wider view of their security and security policies. This does not mean that NATO would become incapable of defending Western European security. Nor does it require the allies to give up the WEU or consultations in the EC.

The US should propose reforming the transatlantic security relationship and eliminating all geographic and functional limits on its competence. The security problems the US sees are all around the world, and include issues confined to specific regions and those that cut across regions, such as the proliferation of weapons of mass destruction. While the world is infinite in the variety of its likely and possible sources of instability and violence, there is a limited pool of partners for the United States to work with to address these dangers. Western European states – Britain and France in particular – have in the past been the primary partners of the United States in responding to threats to their security and that of the United States. Even with reduced capabilities, the best partners for the US in the future will be the Western European allies, and the best way to organise this new transatlantic relationship is through NATO.

The US proposals endorsed at the January 1994 summit for establishing Combined Joint Task Forces and greater allied attention to proliferation of weapons of mass destruction, can be seen as significant, practical steps in this direction.

Under the CJTF concept, task forces would be created within the existing NATO Major Subordinate Commands (MSC) to conduct planning and training for non-Article V missions. NATO's MSCs include AFNORTH (to become AFNORTHWEST in July 1994), AFCENT and AFSOUTH. Non-Article V missions would include all missions not related to defending a NATO member against attack. Peacekeeping or operations in the Persian Gulf would be included in non-Article V missions. The task forces would be able to incorporate forces, staffs and commanders from nations not integrated into the NATO commands (such as France) and would be available to serve under UN, CSCE or WEU mandate. These other organisations could be invited to participate in the staffs on a regular basis or only by liaison. Most important, the task forces would allow European forces to be separable from the NATO structure for the conduct of specific operations without requiring a separate military organisation. This would provide the Europeans with *separable*, but not *separate*, forces. Thus CJTF may be a way forward, allowing the European states to draw on NATO assets as needed. The European states could use CJTF to bridge the gap between policy and the ability to implement mentioned in Chapter II. For the United States, CJTF creates the ability within NATO to plan for deploying forces beyond NATO territory. In both cases, however, European and American, CJTF remains a potential military/operational option for which there is as yet no political consensus.

The proposal for greater NATO efforts against the proliferation of weapons of mass destruction and missiles initiates allied work on the political and defence dimensions of this security threat. The operational details of this initiative will need to be developed, but the need to develop a political consensus is explicit. NATO will develop a political framework for addressing proliferation dangers and begin discussions, including the French, on preventing, reducing and protecting against the threat posed by proliferation. This initiative is extremely significant for the alliance because the proliferation problem presents the potential for a collective response to threats to NATO territory from outside Europe. Moreover, to prevent proliferation before it becomes a direct security threat, the alliance will have to decide whether and how to address collectively the global phenomenon of the spread and transfer of weapons of mass destruction. When North Korea sells missiles to states within range of NATO territory, North Korean actions become a security problem for NATO. Whether and how to respond to this problem is now for the alliance to decide multilaterally and collectively.[6]

This paper argues for a multilateral approach to security in which the US leads and assumes the greatest proportion of risk, but not all of the risk. Even with projected reductions, the US will always be able to field the majority of forces to a coalition and will always be able to claim the leadership and direction of such a coalition. During the Cold War in Europe, the Persian Gulf War and the conflict in Somalia, the United States took the greatest risk and accepted leadership. In Bosnia, it decided to minimise the risk and followed others' lead. In both sets of cases, the US had a multilateral policy, not a unilateral one. The difference, leader or follower, is in the level of US commitment and risk. Under these conditions, whether it leads a multilateral coalition in the future will have more to do with US willingness than European unwillingness. Thus the future of the three recent US initiatives for

NATO – PFP, CJTF and the issue of proliferation – depend on the United States.

The specifics of a wider proposal for remaking NATO follow.

## NATO AND RUSSIA

NATO should approach Russia as a global power and not as any other regional power in Europe. Russia's global status – its continent-spanning size and vast nuclear arsenal – should be the basis for differentiating NATO's relationship with Russia from its dealings with other Eastern states. While Russia should be part of any symbolic extension of Article IV to the NACC states, such as under PFP, NATO's relationship with Russia should not be limited by its work with Eastern Europe because Russia has a greater range of security concerns. NATO should engage in dialogue with Russia on all of its territorial security concerns – European, Caucasian, Central Asian and Far Eastern. This may bring many issues of the Commonwealth of Independent States (CIS) into NATO discussions, such as the handling of nuclear weapons and peacekeeping. By this approach, NATO would emphasise Russia's status as a global power and separate NATO's relationship with it from its ties to Poland, the Czech Republic, Slovakia and Hungary. It may also help to broaden the alliance's traditional view that Europe, and European security, ends at the Urals.

On nuclear issues and the control of weapons of mass destruction, NATO and Russia share a common goal of preventing the further proliferation of these technologies. As part of its new proliferation policy, NATO should establish a new committee on nuclear weapon security and safety open only to those states in NACC that control or host nuclear weapons on their territory. A small subgroup of those states that produce and dismantle weapons (US, Britain, France and Russia) might provide a more conducive working environment for addressing all aspects of handling and eliminating nuclear weapons than larger NATO or NACC groups. It would also give Russia a certain status within NATO. Finally, this would allow Russia to contribute as an equal with the other nuclear powers.

## NATO AND UKRAINE

Ukraine's nuclear weapons, size and enmity with Russia make it an important security issue for NATO's members. Ukraine, while larger than any Eastern European state and having nuclear weapons on its territory, remains a European state and not a global power, and is unlikely to attack NATO intentionally. NATO's security interest lies first in avoiding conflict between Russia and Ukraine, and second in persuading Ukraine to give up its nuclear weapons. Conflict between Ukraine and Russia, especially a nuclear one, might quickly get out of control. Security guarantees for Ukraine from NATO might help convince it to give up its nuclear weapons. Such guarantees, if made credible by NATO members, might also deter Russia from attacking Ukraine. However, NATO is unlikely to offer Ukraine formal security guarantees unless its members are willing to defend Ukraine against Russia and make a policy of deterrence credible. This is even less likely in Ukraine's case than in the case of the Eastern European states that border NATO. The Russia–Ukraine–US agreement on nuclear weapons and Ukraine's territorial integrity signed just after the NATO summit was touted as a great success, and if implemented will be a great boost to European security, but there remains the prospect that a nationalistic parliament will prevent the Ukrainian executive from finally and definitively cashing in the nuclear bargaining chip.

NATO must ensure, however, that Ukraine is not left out of PFP's implementation and that it is offered the full complement of political and mlitary consultations. Responsible civilian democratic government should be the goal of the NATO members *vis-à-vis* Ukraine. As NATO can contribute to this larger question it should, realising that it is a question that goes beyond defence.

NATO AND EASTERN EUROPE
The combination of limited political commitments and extensive military measures discussed earlier under PFP should be pursued as part of a more thorough remaking of NATO. Poland, the Czech Republic, Slovakia and Hungary are a common concern of the alliance, although not a primary one for most members. Because the NATO allies will want to improve their security if possible through a tighter relationship with Eastern Europe, there will be some differentiation in NATO policies towards Eastern Europe and towards Russia. The NATO initiatives with Eastern Europe should exclude nuclear weapons, but be more intensive and at a deeper level of force integration in areas of conventional forces, as described above. Some differentiation is useful both for Eastern European security and for NATO's relationship with Russia.

NATO AND PROLIFERATION
As part of its new mandate for addressing common interests worldwide, NATO should use its review of proliferation policy to place the problem of countering the proliferation of weapons of mass destruction at the centre of its political and military planning process. It

should also begin looking at ways of responding to potentially hostile states that have already begun to develop weapons of mass destruction as part of its primary mission of preventing their further spread. Finally, NATO should use its military planning establishment to develop offensive and defensive military measures for responding to proliferation threats. Again, certain aspects of the proliferation threat, especially prevention measures, would be among the primary topics for discussion with the Russians.

## NATO AND ACTIONS BEYOND ITS TERRITORY

NATO's enforcement of the no-fly zone over Bosnia in 1993 was a significant step towards allowing the alliance to operate beyond the territorial boundaries of its members. This step risks becoming an isolated incident or setting a precedent for only mounting *ad hoc* responses to security problems outside Western Europe. Instead, the step should now be codified in NATO strategy and defence planning requirements. Building on the CJTF, NATO should begin planning for the deployment and sustainment of substantial ground, air and naval combat forces to the Middle East/Persian Gulf and Far East as well as Eastern Europe. Among other missions, these forces might deploy to support UN peacekeeping operations in the Middle East or Africa or to support the UN mandate in Korea should the need arise. This will require a significant restructuring of European defence forces. The cost of developing capabilities for force projection will be tens of billions of dollars. Nevertheless, the cost to Europe will be less than if it attempted to develop these capabilities without the United States, or had to cope alone with its European security problems.

## NATO AND THE SECURITY OF WESTERN EUROPE

The US will remain committed to the security of Western Europe as a core issue of the transatlantic alliance. In order for this commitment to remain credible and viable as the threats to Western Europe diminish, it cannot be the sole purpose of NATO. The development of PFP helps, but in itself is not sufficient as its relevance is also linked to the threat from the East. Western European security, as it is protected by NATO, will continue to be best defended as long as a transatlantic security partnership is the preferred US tool for addressing the most important US security interests. Should Russian reforms fail, and a threat re-emerge, to have allowed the tool of Western security to atrophy would be Western Europe's greatest error. If it would help to reassure the Western European allies of the US commitment to their security, the US could take certain steps and abjure others. NATO's programme of reforming its force and command structure will deepen its forces' multilateral integration through the creation of four multinational main defence corps and a multinational rapid-reaction corps. These units will help to prevent the renationalisation of defence that Europeans fear. Reform of the command structure will make NATO a more European organisation by increasing the proportion of non-US officers in positions of command.

The United States should also consider any political or military arrangement for Western Europe's contribution to NATO. If the Europeans want the WEU to act as an EC subcommittee within NATO, instead of having individual national representatives, this should be acceptable to the United States. It should not matter to the US what organisational arrangement within Europe produces allies for coalitions beyond Western Europe. Provided the Europeans can develop policies that they can implement, spend enough to provide force packages for peacetime integration with US forces and to use in contingencies beyond Western Europe, and are able to react quickly, how all this is accomplished should be a secondary concern. The Europeans, it is clear, want the US role in European security to be managed through and be limited to NATO. The US should be flexible about the institutional relationships of EC states with each other, with NATO, and of the EC and WEU with NATO provided the outcome is effective policy and implementation. Because the US security agenda would be part of the transatlantic deal, however, NATO would have to figure prominently if it is the only institution in which the US can build the diplomatic and military ties needed to make a success of multilateralism.

**Final words**
Provided the recommendations listed above are accepted, NATO and the US role in European security may remain robust and consistent. This is, however, a large package of expensive and politically sensitive issues. To accept it, the Europeans would have to change the way they have thought about the transatlantic relationship and the US contribution to Western European security. PFP and CJTF at their most basic are military/operational initiatives not yet grounded in an acceptance by all of a more fundamental approach to US and European security. Without this grounding in principle and political consensus the ultimate direction and implementation of these initiatives is in doubt. In order to clarify the stakes, the US should make two points.

First, Western Europe must accept and participate in the radical transformation of the transatlantic relationship necessary for continued

80

US support and participation in Western European security. Unless change takes place, the web of political and military relationships that are the alliance, and on which alliance members rely to protect their territorial integrity, will not survive. The status quo is unsustainable.

Second, this new transatlantic relationship will only work if its participants make a strong commitment to multilateral diplomacy and multilateral military action. While no state can be asked to give up its right to defend its interests on its own, the US will make the commitment to bring more of its diplomatic and military initiatives to NATO. The European allies may be dubious of this commitment, but it has the strongest of recommendations: US national self-interest. The ability of the United States to lead is based, first, upon its willingness to do so by force of its own action and incursion of risk. Whether other states follow this lead depends upon their perception of their national interests and their ability to make a contribution to a larger effort. NATO as a multilateral institution can lower the cost of the US taking the lead because it increases the likelihood that the European states will participate with the United States.

For NATO, this agenda means another strategic shift in mission. The shared values of the members of the Alliance remain, but a new strategic review is necessary to encompass the Western European and American security agendas for the future of the organisation. Where the London Declaration linked the security of Western Europe to the rest of Europe, a new document must link transatlantic security to common interests around the world. This link is the best guarantee that Western European security will be assured, even if Russia's trend towards democracy, openness and free markets should reverse and a threat re-emerge.

The United States has said that it would be in Europe as long as the Europeans want it to be there. An essential element of building a consensus in the US for this new transatlantic alliance will be the degree of support from Western Europe. If the US presents this new vision of NATO's purpose as the basis of its engagement in Europe, it will be up to the Western European allies to decide whether they still want the US to play a role in European security under such terms. CJTF is only a tool that may be fully and effectively implemented, left unused or only partially exploited. The new policy on proliferation of weapons of mass destruction and missiles may yield a new mandate for multilateral action or dissolve into arguments over arms exports. The shared values remain, but there may be fewer common interests among the NATO members than there once were. Western Europe may become just one region among many where US interests are at risk, or it may be the home of the most likely and capable allies of the United States, allies the US will need in order to address security outside Western Europe as well as within it. It is clear that the preferred European option for the alliance – the status quo – is not sustainable in the United States. It is not clear whether a US proposal for the defence of common interests would be acceptable in Europe.

# Notes

## Chapter I

[1] Daniel Williams and John M. Goshko, 'A Lesser US Role in the World?: Official's Remarks Bring a Prompt White House Denial', *International Herald Tribune* (*IHT*), 27 May 1993, p. 1.

[2] 'Clinton Pledges Continuity in US Foreign Policy', United States Information Agency (USIA), 4 November 1992; Christopher Testimony to Senate Foreign Relations Committee, US Senate, 13 January 1993; Aspin Testimony to Armed Services Committee, US Senate, 7 January 1993.

[3] Secretary of Defense Dick Cheney, 'Defense Strategy for the 1990s: The Regional Defense Strategy', January 1993, p. 3.

[4] *Ibid.*, p. 4.

[5] 'Bush, Clinton Say Support for UN Will Continue', Text of Statement to UNA–USA, USIA, 9 October 1993.

[6] Secretary of Defense Les Aspin, 'The Bottom-Up Review: Forces For A New Era', 1 September 1993, p. 6.

[7] Christopher Testimony to Senate Foreign Relations Committee.

[8] 'Aspin Outlines New Strategic Partnership', USIA European Wireless File, 5 June 1993.

[9] 'Christopher Urges Collective Action on Proliferation', Text of Statement in Luxembourg, United States Information Service (USIS) Wireless File, 10 June 1993.

[10] *NATO Handbook* (Brussels: NATO Information Service), 1983, p. 14.

[11] See discussion in Chapter II.

[12] Richard G. Lugar, 'NATO: Out of Area or Out of Business', text of speech to Overseas Writers Club, Washington DC, 24 June 1993.

[13] Jim Hoagland, 'The New Agenda: Multilateral Self-Service', *IHT*, 14 June 1993, p. 4.

## Chapter II

[1] Christopher Coker, 'Limited Options, Unlimited Choice?', *European Security Analyst*, no. 23, February 1993, p. 3; Foreign Secretary, Douglas Hurd, Speech to the Diplomatic and Commonwealth Writers' Association, London, 2 June 1992.

[2] Douglas Hurd, Royal United Services Institute, London, 13 October 1992.

[3] Prime Minister Pierre Bérégovoy, Institute for Higher National Defence Studies, 3 September 1992.

[4] Foreign Minister Roland Dumas, WEU Council of Ministers' Meeting, Bonn, 29 October 1991.

[5] Alan Riding, 'France Is Seeking More Global Clout', *IHT*, 28 June 1993, p. 5; Riding, 'France Seeks A Greater Role in NATO', *IHT*, 13 May 1993, p. 6.

[6] *Le Figaro*, 18 May 1993.

[7] Quoted in 'Europe Advances Defense Unity: French Government Might Align Military Closer to NATO', *Defense News*, 24–30 May 1993, p. 4.

[8] Foreign Minister Alain Juppé, Speech before WEU Ministerial Session, Luxembourg, 5 April 1993.

[9] Prime Minister Eduard Balladur, 'European Security Pact Initiative', Copenhagen, 21 June 1993.

[10] 'A Bout of Nerves', *The Economist*, 16 January 1993, pp. 34–35.

[11] *Le Figaro*, 18 May 1993.

[12] 'German Policy in Europe', *Financial Times*, 26 October 1992, p. 2.

[13] Dr Helmut Kohl, Speech at 30th Anniversary of Franco-German Partnership, Bonn, 21 January 1993.

[14] 'Treaty on the Final Settlement with Respect to Germany', *Documents of German Unity* (Hamburg: Friedrich Reinecke Verlag, 1990), Article 6.

[15] General Klaus Naumann, 'Konturen einer gewandelten deutschen Sicherheitspolitik und ihre transatlantische Einbindung', Munich, 7 December 1992.

[16] 'Treaty on the Final Settlement with Respect to Germany', Article 5.

[17] 'Kohl to Order a Sharp Cut in Germany's Armed Forces', *IHT*, 8 February 1993, p. 1. 'Multinational

Forces – Training and Exercises',
*NATO's Sixteen Nations*, no. 1, 1993,
p. 27.
[18] Walter B. Slocombe, Principal
Deputy-Under Secretary of Defense for
Policy, 'The Future of NATO', Senate
Committee on Armed Services,
Subcommittee on Coalition Defense
and Reinforcing Forces, 18 June 1993.
[19] 'Rome Declaration on Peace and
Cooperation', North Atlantic Council, 8
November 1991, para. 6.
[20] *NATO Review*, February 1993, p. 33.
This shows the full drop from the end
of the Cold War and avoids any added
costs of the Gulf War.
[21] International Institute for Strategic
Studies, *The Military Balance 1990–
1991* (London: Brassey's for the IISS,
1990).
[22] 'German Decision to Cut Military
Leads to Unease at Home, Abroad',
*Wall Street Journal*, 17 February 1993,
p. 10.
[23] 'Warning on German Forces',
*Financial Times*, 24 March 1993, p. 3.
[24] Foreign Secretary Douglas Hurd,
Royal United Services Institute,
London, 13 October 1992.
[25] Secretary of State for Defence,
'Defending Our Future: Statement on
Defence Estimates 1993', Ministry of
Defence, July 1993.
[26] *NATO Review*, February 1993.
[27] 'Defense Secretary Details Defense
Expenditure Plans', MoD Press
Release, 30 November 1993.
[28] 'Statement on Defence Estimates
1993', paras 704, 743; *The Military
Balance 1991–1992*, pp. 75–79.
[29] *Ibid.* 'Statement on Defence Estimates 1993', pp. 89 and 99.
[30] *Ibid.*, para. 115.
[31] *Ibid.*, paras 114–22.
[32] *Financial Times*, 5 February 1993, p.
5.
[33] Prime Minister Eduard Balladur,
Before the National Assembly, Paris 8
April 1993.
[34] Philip H. Gordon, 'French Security

Policy After the Cold War: Continuity,
Change, and Implications for the
United States', R-4229-A (Santa
Monica, CA: RAND, 1992), pp. 33–42.
[35] For a more thorough and extensive
discussion of European capability for
force projection, see Richard L. Kugler,
'US–West European Cooperation In
Out-Of-Area Military Operations:
Problems and Prospects' (Santa
Monica, CA: RAND, 1993).
[36] Gerald Turbe, 'The Western European Union and Arms Control',
*International Defense Review*, 'Defense
'92', p. 84.
[37] 'WEU's Satellite System May Fly In
2000', *Defense News*, 1–7 February
1993, p. 4.
[38] 'Anti-Missile Defence for Europe',
Letter from the Assembly, no. 15, June
1993, p. 4.
[39] 'Petersberg Declaration', WEU
Council of Ministers, Bonn, 19 June
1992, part II, p. 2.
[40] WEU Council of Ministers, Rome, 19
May 1993, Communiqué, paras 11–12.
[41] *The Military Balance, 1992–1993*.
[42] M.B. Berman, G.M. Carter *et al.*,
'The Independent European Force:
Costs of Independence' (Santa Monica,
CA: RAND, 1993), pp. 41–42.

**Chapter III**
[1] *The Military Balance*, 1966 to 1975.
[2] 'Defense Strategy for the 1990s: The
Regional Defense Strategy', p. 1.
[3] 'The Bottom-Up Review', p. 6.
[4] Michael R. Gordon, 'Cuts Force
Review of War Strategies', *New York
Times*, 30 May 1993, p. 1; Les Aspin,
'US Reviewing Responses to Military
Threats', Address to National Defense
University, 16 June 1993; '"Win-Hold-
Win" is the New Catch Phrase', *Inside
the Army*, 24 May 1993, p. 1.
[5] 'Bottom-Up Review', p. 6.
[6] Richard L. Kugler, *The Future US
Military Presence in Europe*, R-4194
EUCOM/NA (Santa Monica, CA:
RAND, 1992), p. 22.

[7] Kugler, 'Commitment to Purpose: How Alliance Partnership Won the Cold War', MR-190-RC/FF (Santa Monica, CA: RAND, 1993).

[8] Slocombe, 'The Future of NATO', p. 5.

[9] 'Bottom-Up Review', p. 6.

[10] US Department of Defense, 'Conduct of the Persian Gulf War', April 1992, pp. 319–25, E16–E28.

[11] See discussion on speed of deployment in Chapter III for full notes on this topic.

[12] Richard Sharpe, RN (ed.), *Jane's Fighting Ships, 1993–1994* (Coulsdon, Surrey: Jane's, 1993), pp. 799–800.

[13] 'Bottom-Up Review', p. 11.

[14] Quoted in James Kitfield and Don Ward, 'The Drawdown Deepens', *Government Executive*, May 1993, pp. 46ff.

[15] Barton Gellman, 'Army's Manual: The New Fighting Word', *IHT*, 16 June 1993, p. 2.

[16] General Gordon R. Sullivan, 'Ready for Action: The New United States Army', *NATO's Sixteen Nations*, no. 1, 1993, Special Issue, pp. 62–66. Interview with General Sullivan, 'US Army 1993: Power Projected, Contingency Oriented', *ARMY*, April 1993, pp. 18ff.

[17] For thorough analysis of these options, see RAND report.

[18] Quoted in Kitfield and Ward, 'The Drawdown Deepens'.

[19] Lt-Gen. Helge Hansen, 'Multinational Forces – Training and Exercises', *NATO's Sixteen Nations*, no. 1, 1993, p. 30.

[20] Barton Gellman, 'Aspin Vows to Save US Combat Edge', *IHT*, 21 May 1993, p. 3.

[21] John W. R. Taylor (ed.), *Jane's All the World's Aircraft, 1984–85* (London: Jane's, 1984), p. 439.

[22] William Flannery, 'McDonnell Accepts Pentagon Demands on C-17', *St. Louis Post-Dispatch*, 21 December 1993, p. 6-c.

[23] The SL-7s are container ships originally bought from the US merchant marine fleet, thereby lowering their cost even further. *Jane's Fighting Ships, 1987–1988*, p. 795.

[24] US Department of Defense, 'Conduct of the Persian Gulf War', p. E29.

[25] *The Military Balance, 1992–1993*, p. 25.

[26] Kugler, *The Future US Military Presence in Europe*, pp. 100–01.

[27] John M. Collins, *Desert Shield and Desert Storm: Implications for Future US Force Requirements* (Washington DC: Congressional Research Service, Library of Congress, 1991), p. 10. Collins reports a figure of 'about a month' for sealift to the Gulf.

[28] Kugler, *The Future US Military Presence in Europe*, pp. 100–01.

[29] *Ibid.*

[30] Peter Almond, 'US Floating Force Takes Tanks to Sea', *Daily Telegraph*, 15 September 1993, p. 13.

[31] Paul F. Horvitz, 'House Speaker Sets Conditions for Any US Military Action', *IHT*, 10 May 1993, p. 4.

## Chapter IV

[1] 'Report of the Secretary of Defense to the President and the Congress', January 1993, p. 143.

[2] 'Aspin Outlines '94 Clinton Defense Plan', USIA, 30 March 1993; Barton Gellman, 'Aspin Orders $10.8 Billion Pentagon Cut for '94 by Monday', *IHT*, 5 February 1993, p. 3; Gellman, '"Treading Water", Pentagon Puts Off Key Budget Issues', *IHT*, 29 March 1993, p. 4; Michael Gordon, 'Making the Easy Military Cuts', *New York Times*, 28 March 1993, p. 22.

[3] 'Aspin Outlines '94 Clinton Defense Plan'; Gellman, 'Aspin Vows to Save US Combat Edge', p. 3.

[4] Kitfield and Ward, 'The Drawdown Deepens', pp. 46ff; John Lancaster, 'An Incredible Shrinking Navy', *IHT*, 10 May 1993, p. 3; Barton Gellman, 'Clinton's 1994 Defense Budget, Out

Today, Meets Goal for Cuts', *Washington Post*, 27 March 1993, p. 9; Barton Gellman and John Lancaster, 'US May Drop 2-War Preparedness', *Washington Post*, 17 June 1993; Eric Schmitt, 'Pentagon Is Ready With a Plan for a Leaner, Versatile Military', *New York Times*, 12 June 1993, p. 1; 'Aspin's FY94 Budget Lacks Bite', *Armed Forces Journal International*, 18 May 1993, pp. 17ff; Barton Gellman, 'Aspin Plan Could Cost $20 Billion More', *IHT*, 14–15 August 1993, p. 3; 'Report of the Secretary of Defense to the Congress', January 1993; 'The Bottom-Up Review'.

[5] 'National Military Strategy of the United States', January 1992, p. 19.

[6] *Ibid.*

[7] Warren Christopher, 'Need for American Leadership Said Greater than Ever', transcript of interview in Minneapolis, USIA, European Wireless File, 27 May 1993.

[8] USIS, 'Wireless File 42', 24 March 1992. Quoted in Maynard W. Glitman, 'The Justification for Stationing American Force in Europe 1945–1992', unpublished text.

[9] General John Galvin, 'Hearings before the Committee on Armed Services', United States Senate, 102nd Congress, 2nd Session, 3 March 1992, pp. 382–84. Because the added cost for stationing forces is related in part to local costs, it tends to fluctuate with the exchange rate of the US dollar versus the German Mark, the primary host for US forces in Europe.

[10] *The Military Balance 1992–1993*.

[11] Glenn F. Bunting, 'Citing Jobs, California Delegation Makes a Final Plea to Save 15 Bases', *Los Angeles Times*, 15 June 1993, p. B-1; 'Californian Jobs: Mexico Ho!', *The Economist*, 17 July 1993, p. 42.

[12] Joan Didion, 'Trouble In Lakewood', *The New Yorker*, 26 July 1993, pp. 46–65.

[13] Eric Schmitt, 'A Mission Accomplished', *New York Times*, 29 June 1993, p. 10.

[14] Bruce Carey, 'US Returning More Bases to Host Countries', USIA, European Wireless File, 2 July 1993.

[15] George Graham and Lionel Barber, 'US Should Not Intervene Alone in Bosnia, Says Clinton', *Financial Times*, 25–26 April 1993, p. 2.

[16] Dian McDonald, 'US "Doing All It Can" to Resolve Bosnia Situation', USIA, European Wireless File, 22 July 1993.

[17] 'Allies Remain Hesitant Over Air Strikes', *Financial Times*, 6 August 1993, p. 2.

[18] McDonald, 'US "Doing All It Can"'.

[19] Quoted in Paul F. Horvitz, 'Allies Defend Bosnia Plan, Hinting at Tougher Steps', *IHT*, 25 May 1993, p. 1.

[20] Quoted in Theresa Hitchens, 'Bosnia's Fallout Threatens NATO', *Defense News*, 31 May–6 June 93, p. 1.

[21] *Ibid.*

[22] House Resolution 1621, 103rd Congress, 1st Session.

[23] Lugar, 'NATO: Out of Area or Out of Business'.

**Chapter V**

[1] WEU Declaration, December 1991, para. 2; Maastrict Treaty, Article J.4, para. 2.

[2] This, of course, gives primary credit for US leadership to the greatness of Western Europe. Private interviews.

[3] 'London Declaration on a Transformed North Atlantic Alliance', 6 July 1990.

[4] 'The Alliance's Strategic Concept', 8 November 1991.

[5] 'Christopher Speaks to Press After NAC Meeting', transcript of News Conference, USIA, European Wireless File, 11 June 1993.

[6] 'Declaration of the Heads of State and Government Participating in the Meeting of the North Atlantic Council', Brussels, 11 January 1994.